Cloud Network Management

W0018698

Cloud Network Management

An IoT Based Framework

Edited by
Sanjay Kumar Biswash
Sourav Kanti Addya

CRC Press
Taylor & Francis Group
Boca Raton London New York

CRC Press is an imprint of the
Taylor & Francis Group, an **informa** business

A CHAPMAN & HALL BOOK

First edition published 2021
by CRC Press
6000 Broken Sound Parkway NW, Suite 300, Boca Raton, FL 33487-2742

and by CRC Press
4 Park Square, Milton Park, Abingdon, Oxon OX14 4RN

© 2021 Taylor & Francis Group, LLC
CRC Press is an imprint of Taylor & Francis Group, an Informa business

No claim to original U.S. Government works

ISBN 13: 978-0-367-25605-0 (hbk)
ISBN 13: 978-0-367-55698-3 (pbk)
ISBN 13: 978-0-429-28863-0 (ebk)

Typeset in Epigrafica-Regular
by Nova Techset Private Limited, Bengaluru & Chennai, India

Visit the Taylor & Francis Web site at
http://www.taylorandfrancis.com

and the CRC Press Web site at
http://www.crcpress.com

*To those Network RESEARCHER(s)
whose contribution and sacrifice made the Globe size
smaller.*

Contents

II Standards and Protocol   47

4 Security in Cloud-Based IoT   49

Monjur Ahmed and Nurul I Sarkar

5 Cloud Enabled Body Area Network   67

Anupam Pattanayak and Subhasish Dhal

9   Interoperability and Information-Sharing Paradigm for IoT-Enabled Healthcare   151

Brian Desnoyers, Kendall Weistroffer, Jenna Hallapy,
and Sandeep Pisharody

10 Cloud Computing Based Intelligent Healthcare System   175

Sunita Pattanayak

Foreword

Efficient and secure cloud management of the data gathered by the ever-increasing number of sensors in IoT (Internet of Things) paradigm is crucial for the operational success of many different key applied scenarios, such as smart cities, industry 4.0, precision agriculture, and digital health, to name a few. In this context, the book "Cloud Network Management: An IoT Based Framework," edited by Sanjay Kumar Biswash and Sourav Kanti Addya, is very welcome since it brings contributions from prestigious institutions from different parts of the world, such as MIT, Virginia Tech, SUNY Buffalo, and Auckland University of Technology, among others. The book content should be thus of great value for those interested in getting knowledge about the latest scientific and technological advances in cloud network management in the support for IoT applied scenarios.

Dr. Artur Ziviani
Senior Researcher,
National Laboratory for Scientific Computing (LNCC)
Brazil

Preface

The data storage, processing, and management at a remote location over dynamic networks is the most challenging task in cloud computing and associated networks. Users' expectations are very high for data accuracy, reliability, accessibility, and availability in a pervasive cloud-based environment. It was the core motivation for the Cloud Networks Internet-of-Things (CN-IoT). As mobile users expect to retrieve the personal information arrays with a high degree of mobility support need a unique management methodology for growing information and manipulation procedure for a large data set over the Internet. It is the major challenge in a network-based cloud system. The exponential growth of the networks and data management in CN-IoT must be implemented in fast-growing service sectors such as logistic and enterprise management, where the "Big Data" has a crucial role. The network-based IoT works as a bridge to fill the gap between information technology and cloud networks, where data is easily accessible and available. This book provides a framework for the next generation of cloud networks; it is the emerging part of fifth-generation (5G) partnership projects. This contributed book has the following salient features, i.e., On-demand cloud networks and data access via Internet and digitalization of contexts, Creating a pervasive access network to enable a fully virtual mobile and interconnected distributed environment, and Providing a heterogeneous massive data connectivity in a distributed framework. It achieves higher speed, increased capacity, decreased latency, and better quality of service (QoS) and quality of experience (QoE). It is one of the efficient and promising technologies solutions to meet the high demand for cloud-based network support, it allows short-range, low-power, and low-cost based Internet-based universal data access. The automatic identification of CN-IoT based data handling capabilities allows improving individual data user capability. The large level of implementation has an advantage of cost-efficient data access benefits.

Editors

Sanjay Kumar Biswash received his PhD from the Indian Institute of Technology (Indian School of Mines), Dhanbad, in 2012. Dr. Biswash is currently working as a Assistant professor at NIIT University, Neemrana, India. He held Research Scientist position in Department of Computer Science and Robotics, Institute of Cybernetics, National Research Tomsk Polytechnic University, Russia. He was a Post-Doctoral fellow at the National Laboratory of Scientific Computing (LNCC), RJ Brazil, and San Diego State University, CA, USA. He was a visiting researcher at the University of Coimbra, Portugal. He served as an Assistant Professor (on contract) at the Motilal Nehru National Institute of Technology, Allahabad, India. His research interests lie in Network Management, Mobility Management, 5G, Device-centric Networks, Wireless and Mobile Networks QoS/QoE, Fog and Edge Networks, and Bio-inspired Networks. He serves as an editorial board member for Wireless Personal Communications, Journal of Mobile and Network Applications and reviewer of many reputable international journals and conferences such as IEEE TPDS, IEEE System, IEEE Networks, Elsevier COMNET, Elsevier JNCA, IEEE ICC, IEEE WCNC, etc.

Sourav Kanti Addya received his PhD in Department of Computer Science & Engineering from National Institute of Technology, Rourkela, India. Dr. Addya currently works as an Assistant professor in Department of Computer Science & Engineering at National Institute of Technology Karnataka, Surathkal, India. Prior to joining NIT Karnataka, he was a Post-Doctoral fellow in Department of Computer Science & Engineering, Indian Institute of Technology, Kharagpur, India. He was a visiting scholar at San Diego State University, CA, USA. He obtained his M. Tech degree with national level GATE scholarship from NIT Rourkela, India and B. Tech from the West Bengal University of Technology. He has several international collaborations and sound volunteer research experience as a reviewer of reputable international journals and is a TPC member of international conferences such as Mobile Networks and Applications, Springer, IEEE Systems Journal, IEEE Transactions on Sustainable Computing, IEEE Transactions on Dependable and Secure

Computing, Computers & Electrical Engineering, Elsevier, COMSNETS 2020, IOTSm 2018, GLOBECOM 2015, etc. His multidisciplinary research and technical interests include Cloud and distributed system, Blockchain, Algorithm design, Computer networks, and Information security. Dr. Addya is a Senior Member of IEEE, member of ACM and life member of Cryptology Research Society of India (CRSI).

Contributors

Monjur Ahmed
Waikato Institute of Technology
Hamilton, New Zealand

Supratik Banerjee
NIIT University
Neemrana, India

Sanjay Kumar Biswash
NIIT University
Neemrana, India

Rahul Chauhan
Graphic Era Hill University
Dehradun, India

Brian Desnoyers
MIT Lincoln Laboratory

Subhasish Dhal
Indian Institute of Information
Technology, Guwahati, India

Umang Garg
Graphic Era Hill University
Dehradun, India

Jenna Hallapy
MIT Lincoln Laboratory

R.C. Joshi
Graphic Era Hill University
Dehradun, India

Shahram Latifi
University of Nevada
Las Vegas

Preeti Mishra
Graphic Era Hill University
Dehradun, India

Tapan Naskar
NIIT University
Neemrana, India

Anupam Pattanayak
Indian Institute of Information
Technology, Guwahati, India

Sunita Pattanayak
University at Buffalo, State University
of New York (SUNY)

Arnab K. Paul
Virginia Tech

Sandeep Pisharody
MIT Lincoln Laboratory

Ashok Kumar Pradhan
SRM University, AP
Amaravati, India

Priyanka
SRM University, AP
Amaravati, India

Yenumula B. Reddy
Grambling State University
Grambling, Louisiana

E.Bhaskara Santhosh
SRM University, AP
Amaravati, India

Nurul I Sarkar
Auckland University of Technology
Auckland, New Zealand

Kendall Weistroffer
MIT Lincoln Laboratory
MA, USA

Anurag Satpathy
National Institute of Technology
Rourkela, Odisha, India

Contributors from across the Globe

Abstract

The wireless network paradigms are growing rapidly; the last decade is the witness of it. Networks' capabilities flourished from third generations (3G) to fifth generation & beyond (5GB) steps with the expatriation of high node mobility. Therefore the mobility management became an integral part of wireless networks. The coupling of wireless networks and mobility management is one of the most essential areas of research and development where it can shift-up to the next level for better quality of service (QoS) and quality of experience (QoE).

To achieve the user's experience in wireless networks, it is associated with cloud computing and the internet of things (IoT) and gradually it moves towards fog and edge computing-based networks. It provides a dynamic association and enhanced mobility management procedure. People are looking forward to adding machine learning (ML) system models for intelligent network behaviors. Traditionally, local nodes collect the required data and process the data and transfer it to the remote cloud and system train machine learning models, with the results. The next possibilities are co-located local devices perform the required training, so local data never needs to be uploaded. Using it we can improve the QoE.

Nowadays IoT plays a vital role in modern wireless networks. Therefore, IoT is shifted to the Internet of Everythings (IoE). This predominance generates many IoT based applications such as Healthcare-technology-enabled IoT networks. This infrastructure consists of many legacy medical sensors, IoT-based personal health devices, and software applications. It generates a huge amount of medical data that need to be processed, correlated and analyzed in near real-time computing. Cloud computing and is used for computing in big data. Unique mobility management is mandatory to serve the dynamic schema where an accurate precision and reliability is needed. Such massive data collection and aggregating with appropriate device/user/system require an effective cloud-based data processing system and computation techniques to secure, as well as reliable integration of enterprise IoT networks with public and private cloud networks, as well as personal-connected health devices. The detailed analysis is thoroughly discussed in this book. The healthcare IoT is very often associated with body area networks (BAN). This technology consists of different sensor nodes implanted in the human body to read and analyze the physiological health information (PHI) parameters such as blood sugar, blood pressure, body temperature, etc. If any PHI reading is beyond the normal range of the corresponding PHI, the medical event is reported instantly. These types of works are in the scope of the book.

The wireless technology is not limited to a small geographical area. It is gaining attention for wide areas such as smart cities and smart homes. It also propels us towards several challenges. The IoT can be a good bribe between smart cities and smart wireless networks. The technical amalgamations are equipped with sensors, microcontrollers, transceivers, provided the unique identity (RFID, Barcode, etc.), wireless connectivity and a suitable protocol stack for a seamless connection over a data network. This book will provide a framework to provide a comprehensive study for smart cities, IoT and cloud-based computing models. But it also carries several research challenges and cutting edge technological improvements including security, scalability, research and methodology and cost-effectiveness that are well elaborated in this book.

As the data is a stored and processed over the cloud, therefore it enables several risks and security aspects. Cloud-based IoT is the one approach to overcome it. The distributed nature of a Cloud-based IoT infrastructure is prone to different threats and vulnerabilities related to technological and human-centric factors as well as strategic decisions in the design and implementation of a Cloud-based IoT. This book deals with all of these solutions and methods.

The baseline of cloud networks is a TCP/IP reference model. The model is old enough, and it leads us to new networks theories such as software-defined networks, information-centric networks (ICN), fog and edge networks. The book helps us to find the limitations of current Internet architecture, data processing, information transmission and highlighting the importance of ICN, as it is used to overcome the limitations of legacy network architecture. The significance of ICN with IoT over the cloud is also well explained in this book.

Part I

Evolution of IoT, Cloud Network and Network Mobility

Chapter 1

Evolution of Cloud-Fog-IoT Interconnection Networks

Anurag Satpathy

Department of CSE, National Institute of Technology, Rourkela, Odisha

The traditional cloud computing architecture built on virtual machine (VM) based offerings is faced with multiple challenges such as failure to provide performance guarantee for bandwidth and propagation delays. This made the platform undesirable for modern applications that are network-intensive and have stringent requirements on tolerable delays. To conquer this issue, infrastructure providers (InPs) started to offer resources in the form of virtual data centres (VDCs), that have multiple interactive VMs distributed geographically with different performance requirements. Although services offered through VDCs brought about benefits in terms of improved services and better user experience, issues such as transmission delays and communication overheads were still persistent that made the platform undesirable for the class of latency sensitive applications. As a solution, fog computing was introduced, which brought cloud services closer to the users and thereby reduced the dependency on centralized cloud. Moreover, with the rapid proliferation of Internet of Things (IoT) devices, servicing latency sensitive applications with cloud at the backbone was not feasible. In order to cater to the requirements of such complex applications an interplay between the three stratum, i.e., IoT, Fog and Cloud, is essential. In this chapter, we discuss the evolution of cloud networks starting from the traditional VM based offerings to a more complex model that involves a fruitful interplay between the stratum. Every stage of evolution posed different challenges of which some were extensively researched and of some that did not receive much attention. We study in detail the issues and discuss possible solutions proposed in the literature to address them. Further, we also discuss the strengths and weaknesses of each solution and highlight areas where open research can be conducted.

1.1   Introduction

The traditional virtualization technology was mainly focused at delivering services in the form of computing and storage resources. Although it was able to overcome the drawbacks of traditional data centres (DCs), such as low resource utilization, higher operational costs and lack of isolation, it failed to provide performance guarantee for bandwidth and propagation delays across DC networks. With the outburst of cloud computing, it emerged as a desirable platform for network-intensive applications such as Hadoop [384]. The performance of such applications is heavily reliant not only on computing and storage resources but also on networking resources on which they are deployed. Hence, traditional VM based services were not adequate enough to meet the desired quality of services (QoS) for network-intensive applications. To overcome such limitations, infrastructure providers (InPs) started to offer resources in the form of virtual data centres (VDCs). A typical VDC request is depicted in Figure 1.1. As can be observed from Figure 1.1, a typical VDC request comprises multiple communicating VMs with performance guaranties on delay experienced. Each component of a VDC, i.e., VM or VL, has different resource demands. Coming back to Figure 1.1, the values corresponding to a VM consist of vCPU, memory and disk image demands, respectively. The values corresponding to the VLs denote link resource demands and maximum tolerable transmission delay, respectively. From an InPs perspective allocation of resources in the form of VDCs is posed with a variety of challenges. In the next section, we discuss in detail such issues and solutions proposed in the literature to tackle them.

Although partitioning physical resources at the DCs into VDCs brought about benefits in terms of improved service guaranties and better user experience, issues such as transmission delays and communication overheads were still persistent [392]. In fact cloud computing was still not a viable platform to service cloud based latency sensitive applications such as connected vehicles, fire detection, fire fighting, smart grid and content delivery [279]. To address this issue *fog computing* was proposed that extends the services of cloud closer to the users. Specifically, *fog computing* enables processing of latency-sensitive, real-time and

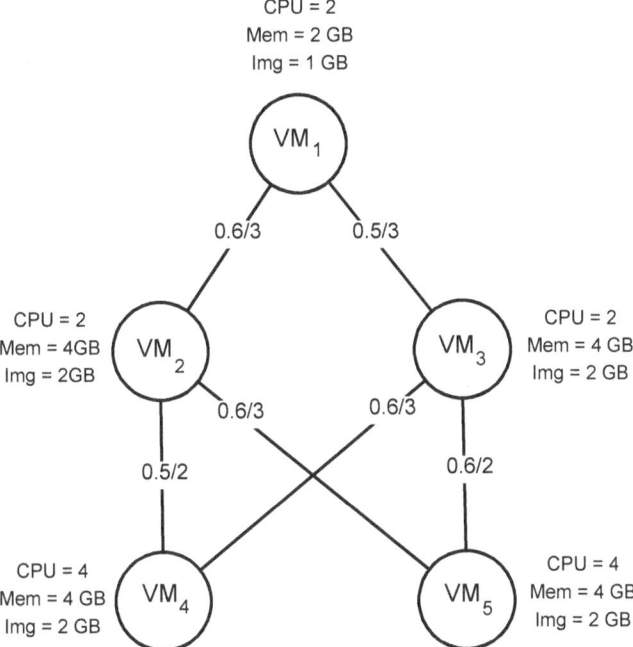

Figure 1.1: A typical virtual data center request.

responsive applications at the network edge whereas latency-tolerant applications can be processed at a distant cloud. In fact *fog computing* is pivotal, particularly when it comes to services supporting data management and analytics.

The rapid growth of Internet of Things (IoT) devices is fuelled by the need for smart objects, including sensors, smart meters, smart cars and actuators, to collect and exchange data to facilitate various applications, such as smart city, smart grid, e-healthcare and home automation [170]. These devices generate huge amounts of data and require almost instantaneous processing, mobility support, geo-distribution in addition to location awareness and low latency that make the traditional cloud an inappropriate platform for such applications. Alternatively, fog computing due to its inherent characteristics can be used as a platform for executing IoT applications [68]. Hence, it can be perceived that rather than cannibalizing cloud computing, fog computing is an enabler to a new breed of applications and services that thrives on a fruitful interplay between the stratum.

Every stage of evolution starting from the traditional VM based offering to VDCs and then to an interplay between fog-cloud and IoT is posed with different challenges of which some were extensively researched and of some that did not receive much attention. We study in detail such issues and discuss possible solutions proposed in the literature to address them. Further, we also discuss the strengths and weaknesses of each solution and highlight areas where open research can be conducted.

The rest of the chapter is organized as follows. The motivation and contributions are highlighted in Section 1.2. Section 1.3 deals with the evolution of traditional VM-based offering to virtual data centres (VDCs) and the challenges associated with it. Section 1.4 discusses the need for fog computing and identifies the challenges associated with it. Section 1.5 discusses the challenges and applications of a more complicated cooperation between IoT-Fog-Cloud. Finally, some conclusions are drawn in Section 1.7.

1.2 Motivation and Contributions

Applications of the modern era not only demand storage and memory but also require networking capabilities. Such applications generally have a strict constraint on acceptable downtime and delay experienced. Departure from requested quality-of-service (QoS) can be perilous for InPs as it can lead to reduced revenue, tarnished market reputation and user dissatisfaction. To mitigate this issue InPs offered resources in the form of VDCs. Moreover, the rapid escalation of IoT devices propelled applications to be deployed at such devices. These devices with limited capabilities were not adequate enough to provide services at real-time; hence, the notion of *fog computing* was introduced. *Fog computing* enabled users to perform processing at the edge of network without relying on the distant cloud. However, the dependency on cloud was not nullified as fog nodes also had limited capabilities in terms of resources. Hence, an interplay between the stratum was necessary that caters to the needs of a variety of applications. Each stage of evolution introduced different challenges of which some were resolved whereas some did not attract much attention. The overall contributions of the chapter can be highlighted as follows:

(a.) To study the evolution of cloud networks starting from VM based allocation to a more complicated infrastructure involving an interplay between the stratum.

(b.) To elaborately discuss the motivation behind such evolution and discuss the roadblocks in the process.

(c.) To perform a systematic review on the challenges faced at each stage of evolution and discuss the techniques used in the literature to address them.

(d.) We also discuss the pros and cons of the solution approaches and identify areas where open research can be conducted.

1.3   Evolution of Traditional cloud networks

As we already discussed in Section 1.1 traditional VM based offering was not suited for many network-intensive, real-time and business critical applications, where virtual data centres (VDCs) are used as a solution. Although evolution of VDCs helped InPs achieve better isolation and service quality, it also brought about a variety of new challenges. VDC embedding/provisioning is one such key problem that InPs have been frequently exposed to. Virtual data center embedding (VDCE) involves mapping/assigning VDC components (VMs and VLs) onto physical resources (servers and physical links) in/across DCs subject to different objectives. The VDCE with constraints on VMs and VLs can be reduced to the NP-Hard multi-way separator problem [98]. Even after embedding VMs mapping VLs is still NP-Hard. A number of researchers have addressed the VDCE problem subject to different agendas such as economic benefits, resource utilization efficiency, energy efficiency, survivability and quality-of-service (QoS) demands. In some cases we will use the terms virtual network (VN) and virtual data centre (VDC) interchangeably as they more or less refer to the same thing with some subtle differences [384].

Since VDCE involves two inter-dependent phases as discussed earlier, Chowdhury *et al.* [98] proposed a VN embedding to coordinate the phases involved. The overall embedding problem was formulated using a mixed integer programming model with substrate network augmentation. The authors devised two embedding algorithms, namely, D-ViNE and R-ViNE using deterministic and randomized rounding techniques. On the other hand, Rabbani *et al.* [310] proposed a VDCE solution considering parameters such as residual bandwidth, server de-fragmentation, communication overhead and load balancing into account. Since energy consumption at the DCs is growing with increasing user-demands, techniques used to reduce energy consumption are crucial to not only increase the revenue of InPs but also reduce its environmental impacts. In this regard, Yang *et al.* [384] proposed two algorithms, NSS-JointSL and NSS-GBFS, to reduce the energy consumption in lightly loaded DCs while minimizing the embedding cost in heavily loaded DCs. On the other hand, Guan *et al.* [147] developed a heuristic based method for VN scheduling considering the expected energy consumption at the DCs and migration costs.

As we have already discussed VDCE/VNE is NP-Hard, which implies that with increasing problem size the solution space and consequently the computation time escalates exponentially. Hence, researchers were motivated to go for meta-heuristic based solutions to obtain a compromised solution in a reasonable amount of time. Ant colony optimization (ACO) is one such meta-heuristic that simulates the behavior of artificial ants that act as agents searching solutions. In this aspect, Guan *et al.* [148] presented an ACO based VN embedding and scheduling technique to reduce the energy consumption of resources across DCs. A link mapping oriented protocol (L-ACS) was presented by Zheng *et al.*

Table 1.1: Summary of Literature on virtual data center embedding.

Work	Provisioning Cost	Acceptance Ratio	Revenue	Single Cloud	Multi-cloud	Energy Consumption	Resource Utilization
Chowdhury et al. [98]	✓	✓	✓	✓	✗	✗	✗
Rabbani et al. [310]	✗	✓	✓	✓	✗	✗	✓
Yang et al. [384]	✓	✓	✓	✓	✗	✓	✗
Guan et al. [147]	✗	✗	✗	✓	✗	✓	✗
Guan et al. [148]	✗	✓	✗	✓	✗	✓	✗
Zheng et al. [165]	✗	✓	✗	✓	✗	✗	✗
Fajjari et al. [126]	✗	✓	✓	✓	✗	✗	✓
Dab et al. [102]	✓	✓	✓	✓	✗	✗	✓
Pathak and Vidyarthi [300]	✗	✓	✓	✗	✗	✗	✓
Sun et al. [341]	✓	✗	✗	✓	✗	✓	✗
Sun et al. [343]	✓	✗	✗	✓	✗	✗	✓
Sun et al. [342]	✓	✓	✗	✓	✗	✗	✗
Amokrane et al. [45]	✗	✓	✓	✓	✗	✓	✗
Dietrich et al. [114]	✓	✓	✗	✗	✓	✗	✗

[165] for mapping VDC requests. Additionally, Fajjari *et al.* [126] proposed VN embedding technique called VNE-AC based on max-min ant systems (MMAS). Genetic algorithm (GA) is also a popular meta-heuristic often used to solve combinatorial optimization problems with a large search space. In this regard, Dab *et al.* [102] proposed a dynamic reconfiguration of VNs based on GA to achieve higher resource utilization and revenue at the InPs. Alternatively, Pathak and Vidyarthi [300] discussed a GA based algorithm to address the issue of VN embedding across a substrate network to maximize the revenue, acceptance rate and resource utilization.

The VDC requests are often exposed to time-varying resource demands which requires reconfiguration of resources. To address the issue of dynamic resource reconfiguration for evolving VDCs, Sun *et al.* [341] proposed an heuristic that aims to minimize the mapping cost and energy consumption for reconfiguring evolving VDC request across multiple DCs. As an alternative, Sun *et al.* [343] have developed a mixed integer linear programming model (MILP) to cater to the demands of evolving VDCs. The authors aim to minimize the remapping cost and resources used. Optimal provisioning of resources for hybrid virtual networks modelled as an integer linear programming model (ILP) to reduce provisioning cost of resources across DCs has been discussed in [342]. Amokrane *et al.* [45] discussed a resource management framework called Greenhead to attain the best possible trade-off between revenue and energy costs for provisioning VDC requests across geographically distributed DCs. Contrasting to the traditional research, Dietrich *et al.* [114] discuss provisioning of VNs across multiple providers with limited information discovery. Table 1.1 highlights and presents a comparative facet of the parameters under consideration for evaluating different strategies discussed in literature.

1.4 Into the Fog

Partitioning DC resources into VDCs improved the service quality and user experience catering to the demands of a new class of network sensitive applications. However, for applications that are latency sensitive, cloud is still not an ideal platform. Some instances of such applications are connected vehicles, fire detection and fire fighting, smart grid and content delivery. The key bottleneck that restricts cloud as an obvious choice for such applications is the poor connectivity between the cloud and end-devices, geo-distributed deployment of applications and service requirements at locations where a provider does not have a DC. As an immediate solution, *fog computing* was proposed that extends services of cloud to the edge of network [280]. This enables the latency-sensitive application to execute their services at the edge and the latency-tolerant applications can be processed at the end cloud. Further, as it is well known that cloud is not an ideal platform for majority of IoT applications, fog could potentially act as the saviour. In fact fog computing acts as a complement to cloud and brings services closer to the users. The fog stratum can be formed by different providers and each domain is formed by a set of fog nodes that include edge routers, switches, gateways, access points, smart phones, set-top boxes, etc.

Fog computing paradigm introduced many new challenges for researchers such as resource management, communication issues and cloud-fog federation. However, it would be incomprehensible to discuss such issues with direct reference to fog as the three paradigms are often interdependent on each other for providing a range of services. Hence, with this motivation, in the next section we discuss the above mentioned issues and their solutions with respect to an interplay between three stratum, viz., IoT, fog and cloud.

1.5 IoT-Fog-Cloud Interplay

In this section, firstly we discuss the challenges that are critical to make an interplay between the three stratum feasible. In this regard, a glaring issue that needs immediate attention is resource allocation and management. We discuss resource allocation and management in three different fronts, namely, resource migrations, allocations and scheduling, and look at different solution approaches proposed in the literature. Further, we also discuss issues such as communication among different stratum and resource allocation in cloud-fog federation. Secondly, we also discuss the application of IoT-fog-cloud interplay for healthcare applications, connected vehicles and smart city applications. All the discussions are conducted with reference to the architecture depicted in Figure 1.2. The lowest layer depicts the end-user strata, middle layer comprise multiple fog nodes and constitutes the fog strata and topmost layer is the high capacity cloud strata.

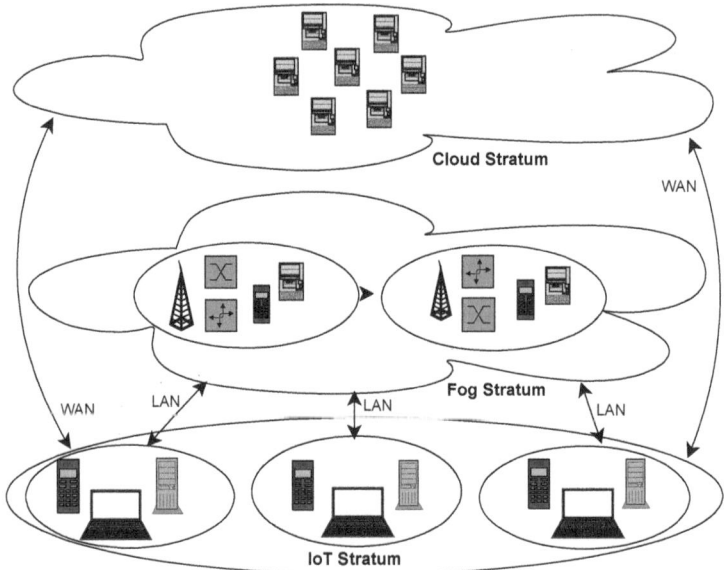

Figure 1.2: IoT-Fog-Cloud Interplay Architecture [279].

1.5.1   Challenges in IoT-Fog-Cloud Interplay

The challenges encountered are discussed mainly in three different fronts, viz., (a.) Resource Management, (b.) Inter- and Intra-Stratum Communication, and (c.) Cloud-Fog Federation.

1.5.1.1   Resource Management

Migrating VMs is an essential aspect of resource management which generally involves seamlessly transferring the state of the VM from one node to another [323], [22]. In this regard, Bittencourt *et al.* [65] discussed a layer based architecture for resource migration focused on VM migration between the fog nodes. The VMs contain user data, components and are available so that as the user moves, the migration is carried out with minimal impact on services. On the other hand, Agarwal *et al.* [23] focused on workload distribution that optimally distributes tasks generated by end-devices (IoT devices) between fog and cloud depending on the availability of resources. Basically, tasks are executed in one of the following three modes, i.e., directly at the fog nodes, suspended and later executed at fog nodes or dispatched to a distant cloud for execution. With regards to scheduling tasks, Cardellini *et al.* [80] focused on exploiting local computing resources at fog nodes keeping the quality-of-services (QoS) requirements intact to schedule data stream processing (DSP) applications. On the other hand, Kapsalis *et al.* [188] departed from the traditional hierarchical and centralized fog model, and adopted a federated model for cooperating edge devices to allocate and manage computational resources required to host varying application components.

1.5.1.2   Inter- and Intra-Stratum Communication

A complex architecture involving three different execution environments, i.e., fog, cloud and local execution (at the devices) often involves tasks constituting an application getting executed at different environments/locations. Hence, proper communication between inter- and intra-stratum is essential and has to be controlled. In this regard, Shi *et al.* [329] discussed an inter-stratum communication protocol between the end-devices and fog nodes. To be specific the authors use constrained application protocol (CoAP) for communication. On the other hand, Krishnan *et al.* [201] proposed an inter-stratum architecture that enables user devices to decide on its destination to execute the task on, i.e., fog or cloud. Contrary to Shi *et al.* [329] that deals with communication between IoT devices and fog, Aazam and Huh [18] studied the communication between fog and cloud stratum and proposed a protocol to reduce the number of packets transmitted to cloud to reduce the overall communication overhead. Alternatively, Slabicki and Grochla [335] explored intra-stratum communication between IoT devices. The authors analyse communication delay for data exchange in three different scenarios, namely, (*i.*) direct communication between devices, (*ii.*) communication through fog, and (*iii.*) communication via cloud.

1.5.1.3   Cloud-Fog Federation

Since, resource sharing is an indispensable aspect of any evolutionary network, hence there is a need to promote sharing of resources between fog and cloud. A major reason that intensifies this cooperation is that fog nodes are resource limited and need assistance of cloud for executing resource-rich applications. Keeping in view such cooperation, Zhanikeev [401] proposed a model called cloud visitation platform (CVP) to facilitate the sharing of resources among cloud and fog stratum.

1.5.2   Applications of IoT-Fog-Cloud Interplay

A interplay among IoT-Fog-Cloud paradigms is quintessential to support modern applications. However, in this subsection we discuss its impacts on applications such as healthcare, connected vehicles and smart city.

1.5.2.1   Healthcare Applications

Healthcare industry has benefited from this interplay architecture as majority of such applications are latency sensitive. Fog computing is an enabler to record and process data of patients' vitals that are monitored by different IoT devices to be stored on remote cloud for future reference. Although not all diseases can be monitored some chronic ailments such as obstructive pulmonary disease (COPD), Parkinson's, speech disorders, and ECG/EEG feature extraction that can be monitored using this architecture. In this context, Stantchev et al. [338] proposed an architecture for nursing services for elderly people. The authors validated their model by a use-case scenario, where IoT devices were used to monitor the blood pressure, fog strata was used for temporary storage of data and cloud was used as a permanent storage to be later referred to by the doctor. On the other hand, Fratu et al. [137] presented a model that extends its support to patients suffering from COPD and mild dementia. In the IoT layer different sensors such as temperature and infra-red movement detectors were deployed. The fog layer was responsible for real-time processing and emergency handling for instance when the oxygen level in the patient's body was out of normal range. Further, cloud is used to maintain data of the patients over a long period of time for future reference. Monteiro et al. [275] proposed a fog computing interface called FIT to analyse the clinical speech data of patients with Parkinson's disease and speech disorders. At the IoT end an android smart watch is used to acquire speech data to be subsequently analysed at the fog nodes and stored at cloud to monitor the progress of patients over a period of time. Gia et al. [141] proposed an IoT healthcare monitoring architecture to detect activities of the heart and brain by exploiting fog and its advantages such as ensured QoS and emergency notifications. Data regarding temperature, ECG and EMG are generated by wearable sensors to be processed at smart gateways acting as fog nodes. Zao et al. [389] developed a brain computer interaction (BCI) game called "EEG Tractor Beam" that monitors the state of the brain using a mobile application on a smart phone. The players of the game are required to identify a ring surrounding a target object and every player has to pull the target towards itself by concentrating. The raw data stream generated during the game is sensed by a smart phone and processed at a nearby fog node.

1.5.2.2   Connected Vehicles

Hou et al. [168] proposed an architecture called vehicular fog computing (VFC) that utilizes vehicles for communication and computation. It utilizes a collaborative multitude of end-user clients or near-user edge devices for better utilization of individual communication and computational resources of each vehicle. Datta et al. [105] discussed an architecture for connected vehicles with Road Side Units (RSUs) and M2M gateways to provide consumer centric services such as data analytic with semantic discovery and management of connected vehicles. Truong et al. [362] proposed a new vehicular-ad-hoc-network (VANET) architecture called FSDN which combines two emergent computing and network paradigms: software defined networking (SDN) and *fog Computing*. SDN-based architecture provides flexibility, scalability, programmability and global knowledge while *fog Computing* offers delay-sensitive and location-awareness services which could satisfy the demands of future VANET architectures.

1.5.2.3 Smart City Applications

In addition to healthcare and vehicular networks the most important aspect of such an interplay architecture is its suitability for smart living and smart cities. In this regard, Li *et al.* [218] discussed an architecture for smart living, i.e., smart healthcare and smart energy. Smart healthcare implies monitoring and detecting chronic heart problems at real-time depending on the data collected by sensors and processed at fog nodes. Concerning smart city applications functionalities of fog nodes can be used for application deployment, network configuration and billing. Yan and Su [380] proposed a smart metering architecture to improve upon the traditional metering scheme. The authors exploit an interplay between the IoT nodes, i.e., the smart home and the smart meters that act as a data node and store user data. Further, periodic data transmission is done to the cloud from fog nodes as a backup. Brzoza-Woch *et al.* [77] designed an architecture for advanced telemetry systems that support automated detection of floods, earthquakes and landslides. This proposal makes use of all three layers with sensors deployed for measurement; many distributed telemetry stations acting for data processing and cloud are used for communication. Tang *et al.* [355] presented a hierarchical distributed fog Computing architecture to support the integration of massive infrastructure components catering to future requirements of smart cities. To add security to future communities, authors propose to build large-scale, geo-spatial sensing networks, perform big data analysis, identify anomalous and hazardous events, and offer optimal responses in real-time using an interplay between the three stratum.

1.6   Research Challenges and Solution Approach

In this section, we identify potential areas where research can be conducted. Considering the network dependency of new age applications and their stochastic resource demands the virtual data centre embedding (VDCE) models can employ learning techniques to predict beforehand the demand spikes and subsequently use a proper relocation mechanism to improve the service quality. The relocation of VDC requests involves VM migration over a distributed network; hence, security is a major concern. The relocations are performed frequently and generally involve transferring of sensitive data; hence, a lightweight cryptographic algorithm should be enforced to provide security for transmission. To solve the security concern, *blockchain* can be used a solution strategy as it enforces a distributed, non-tamperable protocol that is ideal for such environments. Referring to the literature reviewed on *fog computing*, researchers have mainly focused on resource allocation, communication issues and cloud-fog federation. However, the focus has not been on predicting the amount of resources that the fog nodes should posses so that the end-users as well as providers are benefited. The interplay between stratum promotes decentralization and often involves data exchange among nodes that are geographically distributed and has to be carried over a secure channel. This poses a challenge to the traditional cryptographic algorithms that do not support decentralization; hence, *blockchain* can be used a security mechanism.

1.7 Conclusion

In this chapter, we discuss the evolution of traditional cloud architectures based on VM based offerings to virtual data centres that cater to the needs of modern network-intensive applications. However, with the rapid proliferation of IoT devices, servicing latency sensitive and real-time applications using cloud was not feasible. This marked a paradigm shift to fog, i.e., cloud services near to end-users. Since fog nodes are distributed and have limited resources the dependence on cloud for large scale processing and storage was still persistent. In order to conquer this problem, an interplay architecture between IoT-Fog-Cloud was used that not only provides immediate response but also enables large scale processing and storage. Further, such a complex interaction has enormous applications but also introduced many challenges. In this chapter, we discuss some challenges such as resource provisioning, inter- and intra-stratum communication and cloud-fog federation scenarios and highlight approaches used in the literature to handle it. Additionally, we also highlight the applications that can benefit from such an interplay such as healthcare, connected vehicles and smart city applications. Finally, we discuss some future research directions and potential solution approaches.

Chapter 2

Edge or Cloud: What to Choose?

Arnab K. Paul

Virginia Tech, USA

Machine learning (ML) models need to be trained on large volumes of data. Traditionally, client applications collect data and transfer it to cloud servers to train machine learning models, with the results returned to the clients. An alternative to this kind of an architecture is where data is trained at the site where it is collected. Therefore, the data never needs to be uploaded to the cloud server. This is called Edge-based learning, which offers the advantages of privacy preservation and reduced network latency. However, machine learning architects need to be aware of various requirements other than privacy to make an informed decision which to choose: Edge or Cloud? There is a lack of actionable information how these two kinds of learning differ in terms of performance and resource utilization. In this chapter, to address this problem, a comprehensive empirical evaluation of these two types of learning is conducted for a widely used clustering learning algorithm. The results will help in designing learning systems in the future and will help mitigate some challenges faced by ML applications.

2.1   Introduction

IoT (Internet of Things) holds tremendous promise for human society. Current state of the art uses the cloud computing infrastructure as the "brain" for processing and analyzing IoT data, and controlling IoT devices. The low-latency, scalability, and privacy requirements of future IoT applications are motivating the edge computing model: the evolution of the technological landscape that enables in situ data processing and actuation. The data flow in the present scenario consists of IoT devices as the primary data collectors. These devices are generating increasing data deluge, which needs to be operated over to provide applications with high-quality services. Currently, there are two main approaches to operate over the data, one of which is a cloud-based approach in which the collected data is moved across the network to a cloud-based system for processing. Essentially, this approach aims to consolidate the economic utility model with the evolutionary development of many existing approaches and computing technologies, including distributed services, applications, and information infrastructures consisting of pools of computers, networks, and storage resources. The second approach is to train the data closer to the IoT devices called the edge nodes, sending resulting models to a centralized server. The server operates over the aggregated models and updates models on the edge nodes.

There are well known trade-offs between cloud-based and edge-based computation. For example, cloud-based offers additional processing capabilities, with more resources being available at the cloud server. Cloud-based learning may also offer superior values for energy efficiency, as most resource intensive computations are performed at the remote server side, while edge-based offers the decoupling of the model training from the need for direct access to the raw training data. Additionally, the issue of privacy also comes to mind, as the raw data does not leave the confines of the IoT network. Sometimes, latency becomes an important metric to operate over the data. Edge systems will provide lower latency than cloud-based systems.

In order to ease the selection of edge or cloud-based systems for an application, a model is developed in this chapter which takes into account the various factors contributing to the performance of an application. The factors range from system metrics, like CPU and memory to the computation time as well as latency and energy consumption. The model also considers the size of the dataset. The model analyzes multiple runs of machine learning application, namely, KMeans Clustering, and comes up with a decision parameter which optimizes the factors that deem important in the analyses and lets applications ease the decision-making process to decide which system to select cloud-based or edge-based for its computation. Clustering is a machine learning technique that groups items together if they share some characteristics. This technique has been used to solve important problems in diverse domains that include medicine [288, 370, 346], finance [79], social sciences [75], and even search engine optimization [371].

To summarize, this chapter will focus on detailed analysis of the application in both cloud-based and edge-based setup. From the analysis, the optimization model will come up with a list of metrics which are important. Next, based on the inputs from the application, the model comes up with the decision for the application to be built either on cloud or edge.

2.2 Background & Related Work

In this section, the technical background required to understand our conceptual contributions is given.

2.2.1 Edge-Based Learning

Edge-Based Learning is a ML technique where the training dataset remains within the devices which are placed near the source of the data. All devices communicate with each other to compute the model. This ensures privacy preservation, reduced network latency, and less power consumption. The devices can also use the data after it is generated as the data will not be transferred to the cloud. The major disadvantage of this kind of learning is that it is limited by availability of hardware resources [254, 197, 255].

2.2.2 Cloud Computing

Cloud Computing is a distributed computing technique via which resources can be given to applications from a shared pool of resources. The resources can be data storage, compute power or even networks. There is no active management of resources required [260, 135]. This kind of an architecture has major advantages like elasticity, scalability and 'pay-as-you-go' models. For a Cloud-based learning, datasets need to be transferred across networks to the cloud servers thereby increasing risk.

2.2.3   K-Means

K-Means Clustering is a very popular ML algorithm where n observations are divided into k clusters based on which observation is closest to the mean. This reduces the variances within a cluster. An approach to compute K-Means is the iterative technique where, in every iteration, means are calculated based upon a set of points and then after each subsequent iteration, the distance between the means and the points are minimized until they converge.

Work in The Chapter

This chapter conducts an analysis on the implementation of K-Means clustering technique over both Edge-based learning and Cloud-based learning to decipher some key findings for the resource utilization for both approaches. This chapter will hopefully be able to guide ML designers in selecting a system which best suit their needs.

2.3   Experiment

Both edge-based learning and cloud-based learning have been analyzed via a clustering algorithm. The input to the algorithm is a dataset containing geographical coordinates. The output is a set of 4 coordinates, which are basically the four clusters into which all points can be divided to the points with the closest distance.

2.3.1   Edge-Based Learning Procedure

In this kind of learning, client edge nodes collect the geographical data and generate a clustering model that can be then sent to a designated edge node acting as the server. The server combines the models from all the client nodes and gives as output the resultant model. The steps are again repeated for new data.

2.3.2   Cloud-Based Learning Procedure

Here, the process differs from the edge-based learning procedure in the sense that clients are only responsible for collecting geographical data. Once the data is collected, it is sent to the cloud server where the clustering algorithm is used to generate the resultant model on the overall data. This model is then sent back to the clients.

2.3.3 Experimental Objectives

The major objective of the experiments is to compare both learning approaches with respect to system resource utilization, in particular CPU, memory, disk, network and energy. The reason is to understand the system behavior for machine learning algorithms to behave under an edge and a cloud setup. The findings will be extremely useful for both system designers as well as machine learning architects to find the learning procedure which best suits their model.

2.3.4 Setup

The client nodes for both edge-based learning and cloud-based learning are the same. All are Raspberry Pis 3 Model B, with Quad-Core 1.2 GHz and 1 GB RAM running Raspbian OS version3.0.1. There are 9 client devices for each setup. For edge-based learning, the server is located on a Raspberry Pi having the same configuration. In case of the cloud-based learning, the server is a AWS EC2 instance running Ubuntu Server 16.04 LTS, with 8 GB of RAM, with 2 CPU cores. The dataset sizes vary from 100 MB to 3.5 GB. To measure power, a "watts up? Pro" [163] power monitoring device is used. To estimate the power consumption of the AWS instance, energy estimate from Kurpicz et at. [205] is used.

2.4 Analysis

This section has the evaluation results for both edge-based and cloud-based learnings.

2.4.1 CPU Utilization

The CPU utilization (in percentage) is shown in Figures 2.1 and 2.2. It is seen that for clients, CPU utilization is much more for edge-based learning than cloud. The trend is the reverse in servers. This is because, for edge-based learning, individual models are generated in the clients and then passed onto the server, but in cloud the clients are only responsible for data collection.

2.4.2 Memory Utilization

The behavior for memory utilization is similar to that of compute utilization as seen in Figures 2.3 and 2.4. For edge-based learning, memory usage is increased when the model is computed on the clients and then the usage decreases after the model is given to the server, after which the memory usage of server starts increasing. For cloud-based learning, memory usage for clients is high at the

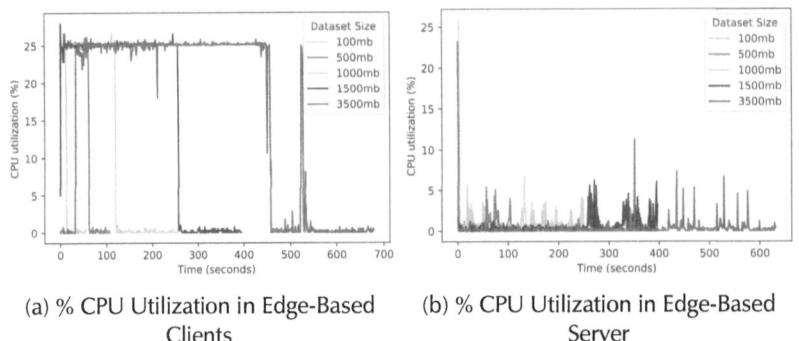

(a) % CPU Utilization in Edge-Based   (b) % CPU Utilization in Edge-Based
                Clients   Server

Figure 2.1: Time-series graphs for CPU %Utilization for Clients and Servers in
Edge-Based Learning

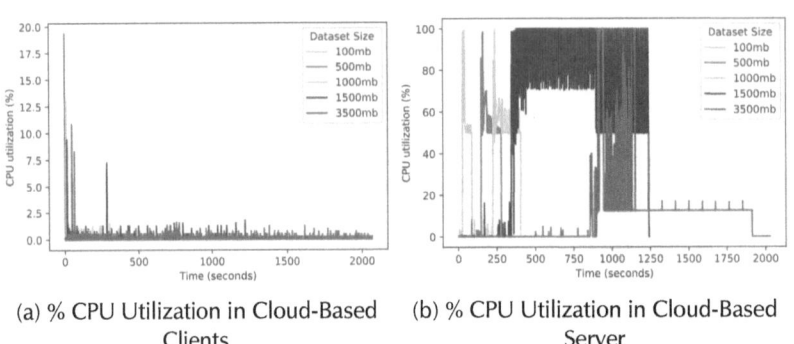

(a) % CPU Utilization in Cloud-Based   (b) % CPU Utilization in Cloud-Based
                Clients   Server

Figure 2.2: Time-series graphs for CPU %Utilization for Clients and Servers in
Cloud-Based Learning

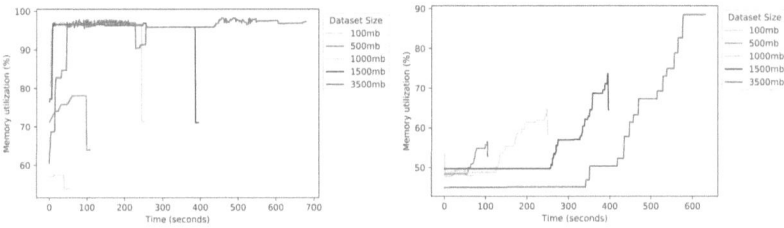

(a) % Memory Utilization in Edge-Based (b) % Memory Utilization in Edge-Based
Clients   Server

Figure 2.3: Time-series graphs for Memory %Utilization for Clients and Servers
in Edge-Based Learning

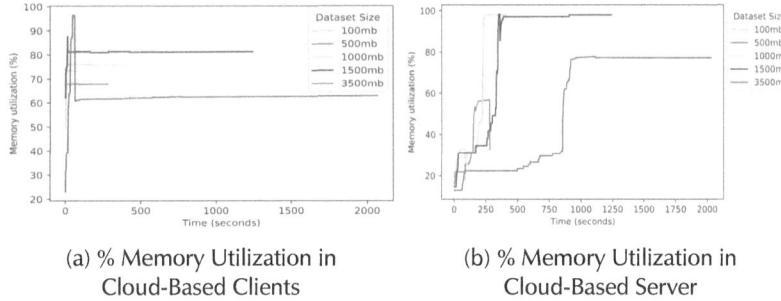

(a) % Memory Utilization in   (b) % Memory Utilization in
Cloud-Based Clients   Cloud-Based Server

Figure 2.4: Time-series graphs for Memory %Utilization for Clients and Servers
in Cloud-Based Learning

beginning when data is collected, after which it gets low and the server's memory utilization increases after that due to the computation of the entire model.

2.4.3 Data Transmission Rate

The data transmission rates in KBytes per second are shown in Figures 2.5 and 2.6. Cloud-based learning achieves much higher rates in data transmission over the network than edge-based learning. Edge-based learning clients have a smaller phase of data transmission over a smaller period of time than cloud-based clients, which are more continuous. This trend is reversed in case of servers where cloud server is much more discreet than edge-based server. But the edge-based server has less data to transmit and therefore the data transmission rates are drastically lower in edge server than the cloud server.

2.4.4 Power Consumption

The time series of power consumption, shown in Figure 2.7, shows that clients in edge-based learning consume more power than in cloud-based learning. This is

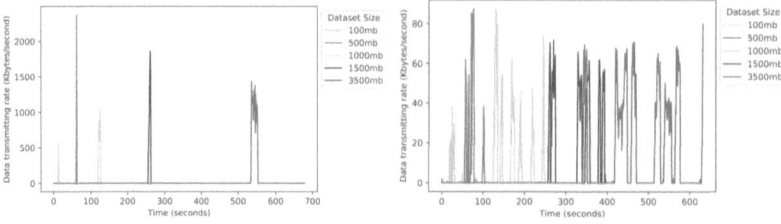

(a) Data Transmission Rate (KB/sec) in   (b) Data Transmission Rate (KB/sec) in
Edge-Based Clients   Edge-Based Server

Figure 2.5: Time-series graphs for Data Transmission Rate (KB/sec) for Clients
and Servers in Edge-Based Learning

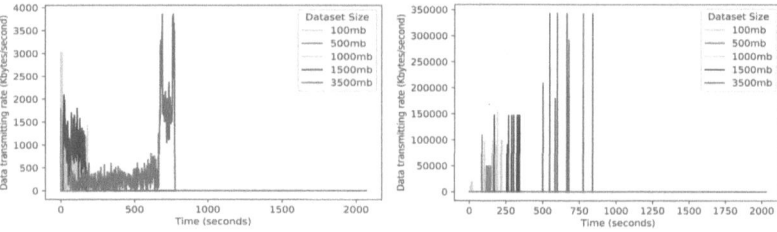

(a) Data Transmission Rate (KB/sec) in   (b) Data Transmission Rate (KB/sec) in
Cloud-Based Clients   Cloud-Based Server

Figure 2.6: Time-series graphs for Data Transmission Rate (KB/sec) for Clients
and Servers in Cloud-Based Learning

because in edge-based learning, clients are responsible for generating the model from the dataset but for cloud-based learning, clients are only responsible for transmitting the data collected.

2.4.5   Energy Consumption

Energy consumption is calculated by multiplying the duration of a job with power consumption. Power consumption was discussed in the previous section. Here, the duration and energy consumption of edge-based and cloud-based learnings are shown in Figures 2.8 and 2.9. As can be seen in the figures, the energy consumption in cloud-based learning is multiple times higher than edge-based learning. This is a very important factor for designers to include when thinking about green computing.

2.4.6   Summary

Edge-based learning consumes less energy and due to the distributed setup of the model generation, also has a lower duration of each job. However, this

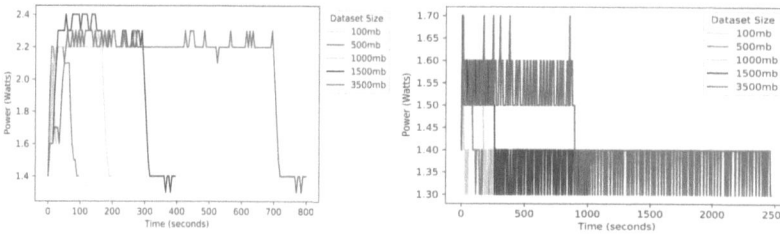

(a) Power Consumption in Edge-Based (b) Power Consumption in Cloud-Based
Clients   Clients

Figure 2.7: Time-series graphs for Power Consumption for Clients in
Edge-Based and Cloud-Based Learnings

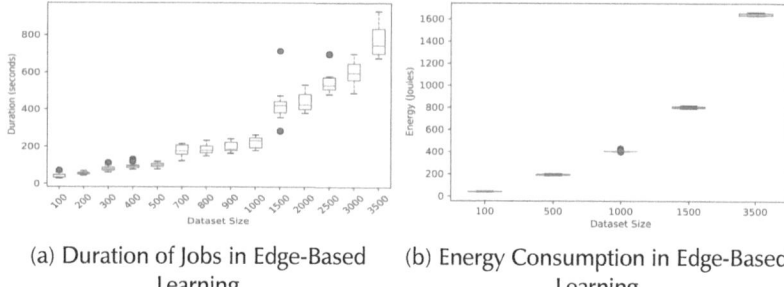

(a) Duration of Jobs in Edge-Based   (b) Energy Consumption in Edge-Based
Learning   Learning

Figure 2.8: Duration and Energy Consumption in Edge-Based Learning

results in higher CPU and memory usage of client edge nodes in edge-based learning than cloud-based learning. Due to AWS network being better than local network, the data transmission rates are higher in cloud-based learning than edge-based learning. However, data transmission needs to happen for a longer period of time in cloud-based learning than edge-based learning due to localized learning.

2.5 Findings

This section focuses on discussing the overall findings for this chapter.

2.5.1 Edge-Based Learning

Edge-based learning acts upon localized data. Clients are responsible for data collection and generating a model for the collected data. This model is then

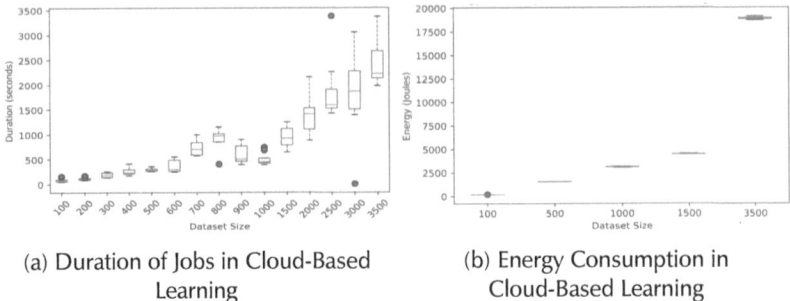

(a) Duration of Jobs in Cloud-Based
Learning

(b) Energy Consumption in
Cloud-Based Learning

Figure 2.9: Duration and Energy Consumption in Cloud-Based Learning

sent to the server for aggregation. The major motivation for edge-based learning is data privacy, as sensitive data will not leave edge nodes where the data is collected. Also, the overall energy consumption and time taken to generate the model is much too low.

2.5.2   Cloud-based Learning

For cloud-based learning, all data is transferred from the data collection client nodes to the centralized cloud server where the machine learning model is generated. Data privacy is hampered in this approach. Also, the overall energy consumption is much higher. However the CPU and memory usage of clients are much less, which will elongate the lifetime of client edge nodes. But this also means that edge nodes are not used to the fullest. Also, data transmission rates are much higher because of an improved network interface in the cloud.

2.5.3   Comparison

The edge-based learning architecture is faster and consumes lesser energy than the cloud-based architecture when performing the same operations over data sets of the same size. However, client devices use more energy in edge-based learning due to model computation at the edge node. System architects and machine learning developers need to take into account the higher resource utilization on the clients for edge-based learning. Another important consideration is the server for the edge-based learning can be of a lower specification because of the lower resource utilization for edge-based learning. Cloud-based learning will incur higher costs due to network management for transmitting data from the clients to the cloud server.

2.6 Conclusion

This chapter focuses on giving a detailed analysis of the two types of learning: edge-based and cloud-based. Both learnings use k-means clustering algorithm to be evaluated. Both offer trade-offs in terms of resource utilization, energy consumption, network latency, data transmission and data privacy. While edge-learning approach helps in lower energy consumption, cloud-based learning helps in lower resource utilization for client nodes. Developers and architects can learn from this analysis to be better informed while selecting an approach for the machine-learning workloads.

Chapter 3

The Survey, Research Challenges, and Opportunities in ICN

Supratik Banerjee

NIIT University, Neemrana, India

Tapan Naskar

NIIT University, Neemrana, India

Sanjay Kumar Biswash

NIIT University, Neemrana, India

This chapter throws light on the limitations of current Internet architecture (TCP/IP) and highlighting the importance of Information-Centric Networking (ICN), as it is used to overcome the limitations of legacy networks architecture. The primary attributes of ICN are the unique naming scheme, in-network caching, routing and forwarding. It is the most usable and best choice for the future Internet architecture. Therefore, the significance of ICN and its association with the Internet of Things (IoT) will be thoroughly discussed in this chapter. We will emphasize the research challenges, real-time development and implementation of ICN-IoT architecture. Finally, we propose the implementation and challenges of edge computing and cloud computing w.r.t. ICN.

3.1   Introduction

The current Internet architecture (TCP/IP) has been designed for host-to-host communication as file transfer was the main design goal of TCP/IP. Thus TCP/IP is ineffectual to adapt to the current challenges i.e., momentous increase in information distribution over the network and variability in Internet traffic. These challenges propelled the idea of Information-Centric Networking[1] (ICN) which proposes the exchange of content[2] rather than communication between hosts and network devices [150].

To improve the availability and performance of the current Internet infrastructure, technologies like Content Distribution Networks and P2P content distribution applications evolved. Both of these technologies are promoting a communication model of accessing data by name, regardless of origin server location. However, their implementation as an overlay adversely impacts the overall efficiency. Moreover, they rely on proprietary distribution technologies. Thus the identification of named information is dependent on the type of distribution channel. Hence, technological innovations to support the identification of named information which is independent of the channel and the source became the need of the hour.

ICN is an approach to evolve the Internet infrastructure to directly support this use by introducing uniquely named content as a core Internet principle. ICN supports data independence from location, application, storage, and means of transportation, enabling in-network caching and replication. Better scalability and improved efficiency are the expected benefits of ICN [67]. ICN also offers a promising solution to problems like security and network congestion. Its

---

[1] The literal meaning of the term "information" is data converted into a more useful or intelligible form. For, eg. host requests for the particular **information**, eg. date-of-birth of Xxx. And he gets in response a text **content** "19-12-1970".

[2] Here the term **content** refers to text, images, audio and video content.

chief design goals like unique naming, multicast communication and in-network caching minimize the response latency and server load.

Moreover, ICN provides promising future for IoT networks because of its unique features like naming mechanism which supports the naming of devices, data, and services for IoT, better security and privacy, in-network caching and in-network processing which benefits the resource constraint IoT devices. And because of decoupling between the sender and receiver in ICN, data can be transmitted in case of intermittent network connectivity of mobile IoT devices [312]. M2M communication in IoT has been already tested on ICN [258]. Moreover, IoT applications like VANET suffer from a frequent disconnection problem which is taken care of by the in-network caching facility in ICN. Thus ICN is envisioned to be an appropriate architecture for modern network schema such as IoT networks [285]. Even though ICN's future is quite promising, its large-scale deployment is facing several challenges like generating name before the content creation in live streaming, updating, and versioning named data objects, accessibility management, privacy protection of requester, data origin authentication, access control, mobility management and implementation of in-network caching.

The rest of the chapter is organized as follows. Section II covers the limitations of current Internet architecture along with its research challenges and issues. Section III covers the basics of ICN, its main features, the prominent ICN architecture, the significance of ICN for IoT, ICN-IoT architecture, the suitability of ICN-IoT architecture for edge computing and cloud computing. Research challenges in the implementation of the same have been also discussed. And Section IV ends the chapter with a brief conclusion.

3.2   Internet architecture and working

The present Internet architecture was designed in the 1960s and 70s when the users had the prime objective of file transfers and internet mail service. The suggested scheme has several technical critiques. Furthermore, the key issue was resource sharing that imposed serious challenges concerning communication among end systems. Thus the current internet architecture(TCP/IP) revolves around communication exchange between hosts and network devices. In TCP/IP Internet hosts can talk to each other by the communication pipes as it is established between them. It serves the client-server application model very efficiently.

With the advent of new technologies primarily at the network core and by the access networks, the bandwidth availability has increased and the user access costs have also reduced. This has resulted in flooding the internet with multimedia applications. And the popularity of multimedia applications with the cost dilution has thronged the Internet with millions of users. Moreover, every year millions of new users are adding up. As predicted by Cisco, over the next 5 years global IP traffic will increase nearly threefold. Remarkably, in 2016 IP video traffic was 73 percent of all consumer internet traffic and by 2021, it will reach 82 percent [17]. As a consequence, applications like video-sharing websites and file-sharing peer-to-peer (P2P) systems evolved. It indicates that the content distribution on the Internet has evolved from a textual information system to a multimedia information system. Thus, the users are more interested in content rather than the content source. Hence, modern applications are content-oriented. But the protocol stack is based on content location and TCP/IP provides no unique solution to this problem. Therefore, the development of network infrastructure that supports content distribution with high availability is the need of the hour. Hence, some partial solutions like Content distribution networks (CDNs) and P2P networks have been proposed for content distribution. And they led the foundation for some widely successful applications like BitTorrent and Akamai. However, both CDNs and P2P systems operate as overlays and the underlying network topology is not taken into account to improve the content distribution efficiency [108].

3.2.1   Research challenges and issues

Following are some of the major research challenges and issues in modern Internet architecture:

- On a global scale, the modern Internet is working as a packet-switched network. And Internet Protocol(IP) is designed to forward the packet using best-effort service. Thus the contents are distributed with no performance guarantee as there is no resource reservation or service differentiation while packet forwarding.

- It compels the users to know the content location as the source host needs to include the IP address of the destination host in the packet header for communication.

- Moreover, the current Internet uses the client-server model in which one point-to-point communication channel is established between one client and one server. Hence, if several users request a particular content hosted by a server, multiple point-to-point channels are needed to be established and one copy of the same content is sent over each of the channels. Thus, popular contents reduce the efficiency of content distribution in terms of bandwidth. Thus scalable forwarding mechanisms are required for large scale content distribution applications.

- In addition to that, the present-day Internet architecture also experiences difficulty in content persistence, scalability, availability, and security. For example, Dynamic redirection in Domain Name Systems (DNS) and Hyper-Text Transfer Protocol (HTTP) is used by CDNs but does not guarantee the persistence of content. Also, the content delivery time is increased as queries to centralized structures are needed to change content location.

• Content authentication and secure communication are also needed for content distribution applications. At present, a secure channel is provided between the source and the destination host instead of securing the content itself. And as a result, additional messages and process overhead are introduced. Moreover, content security depends on the trust of the host that stores the content and also the connection established between hosts.

Thus, mainly three characteristics of current Internet architecture are considered as hurdles to content distribution. There is no guarantee of (i) end-to-end security, (ii) quality of service, and (iii) no scalable forwarding mechanisms.

Hence, modifications are required in the current internet architecture which takes into consideration issues like content delivery, location efficiency, and content availability. And the main goal of Information-centric networks (ICNs) is to fulfill these requirements [108].

3.3   Information-Centric Networks (ICN)

Information-centric networks (ICNs) introduced a paradigm shift in the Internet model. ICN lays stress on content rather than the location which allows the content to be delivered actively by the network. Content naming, name-based routing, in-network caching are some of the innovative concepts employed by ICNs.

3.3.1   Important terminologies used in ICN

This section covers the important terminologies used in ICN:

• **Information-Centric Networking (ICN):** ICN is a concept for communicating in a network that provides accessing named data objects as a first-order service.

• **Named Data Object (NDO):** In ICN, NDO is the terminology used for "the addressable data unit that represents a piece of information or a collection of bytes". And since a name is bound to each data object it is termed as Named Data Object. Different concepts are used in different ICN approaches for mapping Named Data Objects to individual units of transport eg. chunks and segments.

- **Requestor:** In an ICN the entity that is sending the request for an NDO to the network.

- **Publisher:** In ICN the terminology "publisher" is used for that entity that publishes an NDO to the network, and the corresponding request for that NDO sends to that publisher [206].

3.3.2   Concepts and components of Information-Centric Networking

This section covers the general concepts used in ICN which include naming scheme, routing and forwarding, caching.

3.3.2.1   ICN Naming Scheme

In ICN the content/information segment is termed as Named Data Object (NDO). It can be any digital content such as an image, a video, a webpage, a document, etc. And it possesses a name that is location independent. Moreover, data and probably metadata are used for the description of the NDO. The design of NDO varies based on the various approaches used for implementing ICN. Furthermore, Flat and hierarchical naming schemes are used in ICN. However, most recent ICN approaches allow hybrid naming which is a combination of both flat naming and hierarchical naming scheme. And most importantly any of these naming schemes should have the following features:

- **Uniqueness:** It should be assured that content identification is unique.

- **Persistence:** To assure that the content name is unique and valid during the lifetime of the associated content.

- **Scalability:** It should be able to support different namespace scales. Thus, tiny and big namespaces should be served in the same way with no limitation regarding storage location, nature of content or any other characteristics.

3.3.2.2   Routing in ICN

ICN delivers the requested content by its name. And for successful delivery, no information regarding the content source is required. Moreover, information about available contents in the network should be accessible to the nodes. This can efficiently route the valid copies of the content request to the consumer. And this name-based routing strategy needs to have important characteristics like content name based packet addressing which is independent of the source and destination, fault-tolerant routing mechanism, least impact of control information on network traffic and scalable routing. Various mechanisms are there for processing, storing and managing the routing information by the nodes in the network. These mechanisms can be grouped into two: non-hierarchical routing and hierarchical routing.

- **Non-hierarchical routing:** In non-hierarchical routing no structuring mechanism is used to store routing information. And routers are not organized into hierarchical structures. For routing, links are established between the nodes on demand. This allows the nodes to obtain valid content. Since there is no concept of root node for routing information in non-hierarchical routing, the routing information is disseminated among nodes globally. This allows every node to calculate the best route for content delivery. Thus, it provides multiple paths for the same content as the entire topology is known to the node which leads to the trouble-free calculation of loop-free routes. This also increases the availability of the network. As most of the internet routing protocols are non-hierarchical a lot of research work has been done on the same in the past so that it can be applied with some moderation in non-hierarchical ICNs.

- **Hierarchical routing:** In hierarchical routing the routers in the network are connected hierarchically. This ensures that the flow of routing information and data takes place in a deterministic manner. Hence, there is a reduction in the amount of control information as a hierarchical relationship exists between routers. Two types of hierarchical routing techniques are used in ICN: tree-based and distributed hash table (DHT) architectures.

Knowledge of the desired destination node is required in the case of hierarchical tree-based network topologies. Parity, affiliation, superiority, and inferiority are intrinsic to hierarchical structures and it can be applied to name-based routing. Parents nodes have a connection with one or more child nodes. The parent node forms the root of the sub-tree to which the child node belongs. Peer nodes belong to some hierarchical level and have a common root node. Parent nodes accumulate and manage all the routing information of the child nodes. Moreover, the parent node also aggregates the routing load for its entire subtree, which reduces the amount of information used by each child node in routing. And thus the child nodes require less computational power and memory. Any node needs to maintain only the routing information of its parent, peer and child trees. This parent node is the single point of failure which can result in the removal of the entire branches of the tree and thus the content distribution is effected for that entire sub-tree.

Distributed hast tables (DHT) are used for the distribution of cryptographic hash keys among participating nodes. And to ensure protection against a single point of failure, processing and caching costs are shared among nodes. Hierarchical DHTs are used for arranging nodes hierarchically in overlay networks, which helps to efficiently forward the messages towards hash keys [379].

3.3.2.3   In-Network Caching

Internet content access characteristics show that a small number of popular content is responsible for most of the traffic in the network. And network performance can be improved by fetching content from nodes that are placed geographically closer to end-users. Hence, as the content is forwarded to various nodes, most frequently accessed contents can be cached in the router's memory. The in-network caching facility achieves this objective by distributing the copies of contents to distant nodes, closer to the end-users. Named Data Objects in ICN allow caching at any network element including proxy caches, routers, and end-user devices. However, it creates more challenging problems for the routing protocols. Since the content router aggregation process becomes more complicated due to in-network caching, routing information management becomes complicated which impacts the efficiency of the protocol [116].

3.3.3   ICN Architectures

Some of the main ICN architectures are discussed in this section with the primary focus on the Named-data networking architecture which has attracted the focus of most of the researchers.

3.3.3.1   Data-Oriented Network Architecture (DONA)

Data-oriented network architecture is the first ICN architecture which is based on the clean-slate design for persistent and secure content naming, content distribution and name-based routing in a hierarchical network. DONA uses self-certifying flat names which provides persistence and authenticity to content. Content requests are sent to the best serving nodes by the routing mechanism. And the principal or the publisher is responsible for generating content names in DONA architecture. Moreover, each content name is associated with a public-private key pair, which is used for content identification. Automatic server selection, multi-homing, and mobility are intrinsic to DONA architecture [198].

3.3.3.2   Named-data Networking

Named Data Networking (NDN) is one of five projects funded by the U.S. National Science Foundation under its Future Internet Architecture Program. And Content-Centric Networking (CCN) served as the predecessor of NDN.

NDN communication is driven by the receiver, i.e., driven by the data consumers. The communication is realized using two types of packets: Interest and Data. Both Interest and Data packets bear a name that recognizes the content transmitted by the Data packet. The interested consumer attaches the content name in an Interest packet and transmits it. This name is used by the router for forwarding the Interest packet in the direction of data producer(s). Once the Interest packet reaches a node containing the required data, the node returns a Data packet containing the desired content, content-name, and signature by the

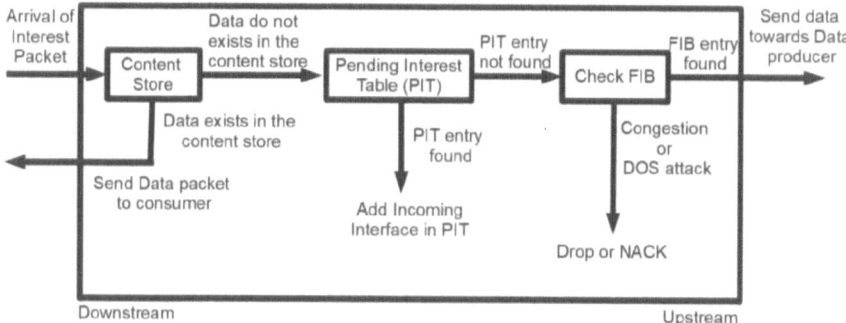

Figure 3.1: Forwarding process of NDN node

producer's key (Figure 3.2). This Data packet follows the reverse path taken by the Interest to get back to the consumer which requested the content.

For forwarding the Interest and Data packets, three data structures are maintained by each NDN router: a Pending Interest Table (PIT), a Forwarding Information Base (FIB), and a Content Store (CS) (Figure 3.1), as well as a Forwarding Strategy module which determines the policy to forward each Interest packet. The PIT stores all unsatisfied Interests forwarded by the router. Moreover, the PIT entry maintains the records of the data name carried on the Internet, along with its incoming and outgoing interfaces. On the arrival of an interest packet, the NDN router needs to check the content store for the matching data; and if the router finds a matching data it is returned to the interface from which the Interest came. In case the router does not find the matching data, the router looks up the name in its PIT, and if a matching entry exists, it records the interface from which the Interest packet was received in its PIT entry. And if no matching PIT entry is found, the Interest is forwarded towards the data producer by the router based on the information from the FIB as well as the router's adaptive forwarding strategy. In case the router receives multiple Interest for the same content from many downstream nodes, only the first one is forwarded by the router upstream towards the producer. The FIB is populated based on a name-prefix based routing protocol, and it can have multiple output interfaces for each prefix.

In special circumstances the forwarding strategy may drop the Interest packet, e.g., congestion in all upstream links or the Interest is suspected to be a DoS attack. For each of the matching Interests, the longest-prefix matched entry is retrieved from the FIB by the forwarding strategy. And then the forwarding strategy decides when and where to forward the Interest. Each router has a temporary cache called content store which is used to cache the Data packets. An NDN packet is meaningful and independent of its source and destination. So an NDN packet can be cached to satisfy future requests.

On the arrival of a data packet, an NDN router searches for the matching PIT entry. And the data is forwarded to all the downstream interfaces listed in that PIT entry. Then the PIT entry is removed by the router, and the Data is

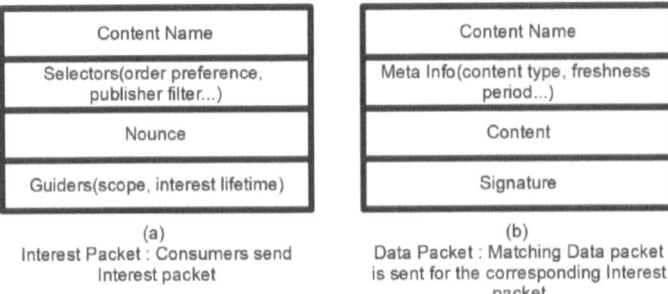

Content Name
Selectors(order preference, publisher filter...)
Nounce
Guiders(scope, interest lifetime)

Content Name
Meta Info(content type, freshness period...)
Content
Signature

(a)
Interest Packet : Consumers send Interest packet

(b)
Data Packet : Matching Data packet is sent for the corresponding Interest packet

Figure 3.2: Format of Interest packet and Data packet in NDN
Reference: Adapted from [397]

cached in the Content Store. The path taken by the Data packets is always the reverse of the Interest packet, and, in the case where there is no packet loss, one Interest packet results in one Data packet on each link, which results in flow balance. Interest or data packets do not carry any host or interface address; the Interest packets are forwarded by the router towards the data producers based on the names carried by the packets. And the data packets are forwarded to the consumers based on the PIT state information set up by the Interests at each hop. This packet exchange symmetry between Interest and Data packets induces a hop-by-hop control loop thus eliminating the requirement for any notion of source or destination nodes in data delivery.

- **Naming:** NDN naming is independent of the network, thus routers do not convey any meaning to the names. It allows each application to decide its naming scheme freely in the network. Furthermore, the hierarchically structured naming scheme used by NDN has its advantages. Applications can easily represent the context and relationships of data elements using a proper naming scheme. For example, an image produced by XXX University may have the name xxx/cse/images/demo.png, where '/' delineates name components in text representations. Segment 2 of version 3 of an xxx video might be xxx/name/videos/vid.mpg/3/2. However, in special cases, flat names can be accommodated which may be useful in the local environment. But, hierarchical names are more useful for providing the necessary context to the data as well as for scaling.

- **Security:** Security in TCP/IP is managed by the endpoints, but NDN security is provided to the data by getting the data packets cryptographically signed by the data producers. The publisher's signature provides data origin authentication and also ensures integrity. Thus, this helps to decouple consumer's trust in data from how and where the data is obtained.

- **Routing and Forwarding:** The routing and forwarding process in NDN is done using names. This eliminates three problems encountered due to addressing in IP architecture: depletion of address space, Network address translation, and address management. There is no address exhaustion problem in NDN since the namespace is unbounded. And since NDN has no public and private addressing scheme there is no need for NAT. So, in the local network address, assignment and management is no longer an issue.

Conventional routing algorithms like link-state and distance vector can be used in NDN. However, instead of announcing IP addresses, an NDN node announces name prefixes that identify the data the node is willing to serve. The routing protocol needs to announce this information across the network, which helps each node to construct its FIB. And other conventional routing protocols like BGP and OSPF can be also adapted to route on name prefixes by performing component-wise longest prefix match of a name in an Interest packet from the FIB table. Moreover, OSPF Based Routing Protocol for Named Data Networking (OSPFN) has been successfully deployed and tested on the NDN testbed [372].

However, Routing Information Protocol (RIP) has been designed to manage a relatively small network. A RIP router broadcasts routing information to its directly connected networks every 30 seconds. ICN is envisioned to connect billions of devices. And the routing information needs to be updated much more frequently as the data is kept at multiple sources due to in-network caching. Thus RIP is not suitable for ICN.

Forwarding across NDN is supported by the Pending Interest Table (PIT). PIT records each pending interest and incoming interfaces, and interest is removed after the receiving of the matching interest or after a timeout. And based on performance measurements and FIB information, the forwarding strategy module in each router makes decisions about which Interests to be forwarded to which interface, how many unsatisfied Interests to be allowed in the PIT, the relative priority of different Interests, load-balancing while forwarding Interest to various interfaces, and in case failures are detected how alternate path should be chosen. However, if the router determines unsatisfied Interest due to reasons like the upstream link is down, no forwarding entry in the FIB or very high congestion, the router sends NACK to its downstream neighbors that transmitted the Interest. After receiving the NACK router forwards the Interest to other interfaces. And the router PIT state identifies and resolves the looping packets.

- **In-Network Storage:** As name and signature are carried by each NDN packet, it is meaningful and independent of who requested it or from where it is retrieved. Thus the future requests can be satisfied by the router if it caches the Data packets. The content store used in in-network caching is synonymous to the buffer memory in IP routers, but IP routers have no facility to reuse the packet after forwarding it to the destination but NDN router can perform the same. Thus, the in-network caching facility minimizes the delay in data delivery. Hence, data delivery latency is significantly minimized in the case of static files. Even dynamic content can benefit from caching in the case of multicast (e.g., realtime teleconferencing) or retransmission after a packet loss.

In addition to the Content Store, NDN architecture supports persistent and large-volume in-network storage facility, named as Repository (Repo for short). This facility is similar to modern days' Content Delivery Networks (CDNs). And no application-layer overlays are required (DNS manipulation) to make it work.

Caching named data raises privacy issues different from those of IP. In IP packet headers can be analyzed to learn who is consuming the data. However, in NDN naming and caching of data may help to check what data is requested, but since no destination address is there in the packet, it is harder to identify the requester. Thus NDN provides different sorts of privacy protection compared to IP networks.

3.3.3.3   Other architectures

There are some other prominent architectures in ICN. Content-Centric Inter-Networking Architecture (CONET) [112] which proposes a new CONET layer that provides users with network access to remote resources that are named. Publish-Subscribe Internet Routing Paradigm (PSIRP) which is based on publish/subscribe based network [115]. And NetInf aims global communication by connecting different technologies and administrative domains under one roof, i.e., ICN.

3.3.4   Information-Centric Networking based Internet-of-Things

3.3.4.1   Why ICN for IoT?

To overcome the shortcomings of IP-based networking, ICN is considered to be a promising approach. Content naming is used by ICN to get rid of address-space scarcity, and the content is accessed by name-based routing. Contents are cached at intermediate nodes to provide reliable and efficient delivery of the data. Moreover, the contents are self-certifying which ensures better security. Thus, the benefits of ICN in terms of data delivery speed, efficiency, and improved reliability paves the way for the ICN based Internet of Things (IoT).

Connecting all the devices with the Internet is the main aim of IoT so that all the devices can be accessed at any time and from any place. IoT paves the way for future technologies like smart washing machines, smart refrigerators, smart microwave ovens, smartphones, smart meters, and smart vehicles. Associating these smart objects with the Internet provides the pathway for remarkable innovations like smart home, smart building, smart transport, digital health, smart grid, and smart cities. And bonding billions of these devices to the Internet leads to the generation of an enormous amount of data which results in IoT Big Data. However, systematized access and discovery of IoT Big Data put more limitations on the underlying TCP/IP architecture while raising many critical issues.

From the IoT perspective, one of the issues is the naming (and addressing) of every IoT device. As the IPv4 addressing space has been exhausted, IPv6 address space may also exhaust in the future. And the long length of IPv6 addresses makes it less suitable for communication through constraint-oriented devices like wireless sensors [327, 69]. Hence, IP architecture does not provide efficient naming and addressing scheme for billions of devices. Moreover, different devices have divergent specifications and constraints in terms of processing power capability, size, memory and battery life which procreate complicated issues. And most of these devices are low power, tiny, low cost, limited memory, and constraint-oriented wireless sensors. And these devices frequently suffer from data unavailability problem. However, IP-based networking provides no solution to this data unavailability problem. But, ICN handles this problem quite smartly using its in-network caching facility. IoT applications like smart home, smart town, smart health, etc., require more security and privacy whereas VANET and MANET require better mobility handling [330, 32].

From the data perspective, most of the users of IoT applications are more interested in getting updated information rather than knowing the source or address of information. For instance, in the domain of wireless sensor networks (WSN), IoT devices aim to harvest information on a large scale [30].

TCP/IP network architecture was designed to connect a limited number of computers to the network and for the sharing of limited network resources. Thus it was not designed for modern IoT requirements. Moreover, the huge amount of data produced by IoT devices put additional requirements like scalability and data dissemination on the underlying architecture. Thus, to fulfill these requirements, ICN can serve as a promising architecture. Its primary characteristics like in-network caching, content naming, better and easy mobility management, improved security and scalable information delivery make it suitable for IoT applications. Additionally, ICN can mask over the TCP/IP network layer or MAC layer. Even though ICN has an edge over TCP/IP, it will take time to completely replace TCP/IP. Thus, shortly ICN will work as an overlay on the IP network. And in the long run, TCP/IP will be completely replaced by ICN architecture.

In-network caching in ICN can efficiently handle information delivery from an unavailable device (i.e., dead device due to the expiry of battery life) by caching contents at intermediate nodes. In-network caching also minimizes the content retrieval latency, provides scalable and easy information retrieval of a large amount of data produced by IoT applications. And better hand-off for mobile devices like mobile phones and vehicles is provided by mobility handling features. Moreover, ICN's feature of self-certifying contents provides better security [41, 136].

3.3.4.2   IoT Architecture Requirements

Following are the requirements and challenges [53, 30, 69] posed by IoT network architecture:

1. Scalability: The main vision of IoT is to connect networks, corresponding devices and billions of low power devices to the Internet. And this

poses bigger and newer challenges over underlying architecture because of scalability issues. IoT architecture should support billions of devices efficiently. IPv6 provides a huge address space to accommodate IoT devices. But addressing is not the only issue. A large amount of data that is being produced by IoT devices needs to be managed efficiently while considering the scalability issues. Therefore, IoT network architecture must be scalable for content access and at the same time it must maintain network efficiency.

2. Mobility: Mobile devices like smartphones, tablets have limited battery life. And some of the IoT applications need connectivity anytime and anywhere. Furthermore, the number of mobile devices connected to the Internet exceeds the stationary nodes. Thus for data availability, reliability, and faster connectivity, the network architecture should provide seamless mobility and roaming.

3. Security and Privacy: In specific IoT scenarios transmitted data is highly sensitive, for eg. smart health and smart hospital. Hacking of such data can lead to severe consequences. Thus, authorization, confidentiality, and integrity should be incorporated in the IoT network. Data access policies and standards should be also clearly defined. For example, a smart home where the detail of a soup ordered by the house owner is required by a hotel for processing payment. But if this detail is shared with his insurance company then this can affect user policy. As the private data can be misused by the insurance company, hence, privacy must be ensured via some access policies.

4. Naming and Addressing: Billions of tiny low-power constraint-oriented devices are used in IoT. These devices need unique naming and addressing to get recognition in the IoT network. Because of the availability of large address space in IPv6, addressing and naming a large number of IoT devices is no longer an issue. However, IPv6 uses long addresses. Hence, it would be complex to process these addresses for constraint oriented IoT devices as it leads to wastage of resources. Moreover, IoT contents are produced and processed very fast. And, for a single content, there are multiple versions. Thus, naming these contents increases complexity. Hence, efficient naming schemes to suit the IoT environment is needed.

5. Heterogeneity and Interoperability: The primary components of IoT are smart wireless sensors and RFID tags. Smart sensors are the major components of IoT and it offers many applications. And since these devices vary in specifications like memory size, processing power, and battery life, thus they are heterogeneous. Moreover different technologies like cellular technology, Bluetooth, 4G, LTE, CRN, and opportunistic networks are used to communicate between these sensors. Hence the technologies used for communication are heterogeneous in nature. Thus the technology required for IoT must support heterogeneity.

6. Data Availability: Whenever a node moves from one location to another in TCP/IP architecture, data may become unavailable. A similar case occurs when a battery completely drains out and is unable to forward data. Similarly, due to the Denial of Service (DoS) attack Internet users cannot receive data. Since TCP/IP architecture cannot look at or inspect data according to request during data transmission, consequently DoS attack is possible. And, the in-network caching technique can improve data availability.

7. Energy Efficiency: A huge amount of energy is needed by billions of devices to build IoT applications. And most of the smart devices such as wireless sensors are low in battery life. Thus, to make universal connectivity possible in the form of IoT, energy-efficient mechanisms are required.

3.3.4.3 Significance of ICN for IoT

IoT users, as well as the internet users, are interested in data and not the data source. And ICN does the same for the users, as it provides the required data to the consumer without knowing about the producer of data. The receiver driven communication model in ICN works on the same principle. In ICN receiver requests for the content and need not bother about the source of content. Moreover, data access is only possible when the receiver explicitly requests it.

IoT plays a crucial role when a vehicle faces a road accident and that vehicle wants to inform incoming vehicles about this accident. In this case, any specific node can act as both the producer and consumer. Hence it results in the flash crowd as only one vehicle is providing the data about the incident. The request for specific information by a large number of users in the network causes the flash crowd. Similar flash crowds are also produced in modern-day internet [24, 379]. Hence, flash crowd results in an increase in network traffic for a particular server [49], which results in quick drainage of the battery life of the sensors of producer vehicles. Thus data can become unavailable due to the end of the battery life of some of the sensors. This problem can be sorted out by the in-network caching facility provided by ICN as it minimizes the traffic load on the original data producing server while caching the data on intermediate routers. And these intermediate routers can provide the cached data on behalf of the original producer thus reducing the so-called flash crowd situation. Thus, the in-network caching facility serves as a boon for the low powered IoT devices. And since the content is named independent from its source in ICN it can be stored anywhere globally. Moreover, ICN provides names to content; thus it is more suitable for IoT as it can combine billions of devices and huge information contents. Moreover, for push type communication beacon messages are used [244]. As the content name is location independent, opaque communication is provided by ICN between the sender and receiver making it more secure.

3.3.4.4   IoT Requirements Mapping to ICN Characteristics

Scalability requirements for IoT applications that connect to billions of IoT devices and produce a massive quantity of content can be fulfilled by ICN characteristics like content naming, in-network caching and content-based security. ICN naming and name resolution can be efficiently used to provide billions of addresses and names to IoT devices and contents respectively.

Receiver-driven communication in ICN along with flexible naming and location independence makes hand-off easier for mobile devices in IoT applications. Moreover, ICN in decoupled mode can perform easy re-registration after a hand-off of a mobile device with the nearest new router [50].

Security and privacy in IoTs can be provided through the following features of ICN.

1. ICN content naming made it easy to inspect whether data is flowing according to query or not.

2. Content location independence hides the source of content.

3. Receiver-driven communication style confirms that content arrives because the receiver has requested for this content.

4. Self-certified contents ensure that the contents are the same as sent by the source.

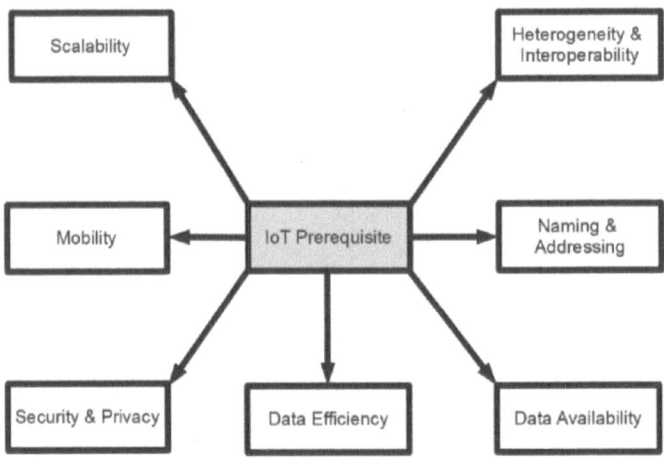

Figure 3.3: IoT prerequisites

Heterogeneity among IoT devices can be easily handled when devices are named using ICN naming scheme. Different types of IoT devices can operate with each other more efficiently when ICN strategy layer is induced in IoT devices. Data availability in the IoT network can be improved by the ICN in-network caching facility. Moreover, in-network caching decreases the frequency of fetching data from the producer and thus saving network life and making it more energy-efficient. Figure 3.3 shows the summary of IoT requirements.

3.3.4.5 ICN-IoT network architectures

ICN-IoT architectures were proposed to fulfill the needs of IoT infrastructure. On the same lines research efforts have been discussed in this section. And the significance of this research area and the efficient applicability of ICN for IoT has been also discussed.

The research community is actively working on IoT to suit the needs of ICN architecture. Thus NDN based ICN architecture was proposed for IoT [40]. In NDN-IoT architecture there are three layers, namely, the application layer, the NDN layer, and the thing layer. To enable name-based networking, architecture consists of content chunks instead of IP addresses. Moreover, to support the transport and forwarding tasks strategy layer is introduced. NDN operates at the network layer and performs its task with the help of two planes, namely, control and management plane and data plane. Tasks like routing, configuration and service models are performed by control and management plane. While the data plane manages the Interest and Data messages and the caching strategy.

There are three different strategies in NDN to support push-type communication so that IoT push operations can be supported [42]. Primarily NDN is designed for pull-based communication but to provide support for IoT push support has been introduced in IoT. In the first scheme called Interest notification, the Interest message is modified by including small data that is to be transmitted. And this data is not cached. The second scheme called Unsolicited data transmits a small packet of uData. And in the third scheme called virtual interest polling (VIP), long live interests are transmitted by the receiver and whenever data is available producer replies and in case of failure consumers can re-transmit an Interest message. All these three techniques have their own set of advantages. In terms of network resource usage, VIP outperforms the other two and is suitable for a massive IoT environment. But Interest notification and Unsolicited data techniques are suitable where battery life is important.

To provide scalability to IoT, CCN(NDN) is the best candidate and it has been implemented in RIOT OS through simulations [57]. ICN has been deployed and implemented using 60 nodes located in several rooms of several buildings. CCN-lite is a lightweight version of CCN which simulated and enhanced CCN through two routing protocols, namely, vanilla interest flooding (VIF) and reactive optimistic name-based routing (RONR). Both VIF and RONR reduce routing overhead for constraint oriented devices. Moreover, the positive aspects of caching and naming the data have been also addressed. And the security aspect of the NDN-architecture has been implemented in Python. And Javascripting-based browser has been used for data visualization. The same has been installed at UCLA (the University of California at Los Angeles) [326]. Access control methods which are Name-based and encryption-based have been proposed to secure sensitive data.

To showcase the scalability and security performance achieved by NDN this initial prototype has been developed. And to address the issue of heterogeneity in IoT for both static and mobile devices a unified ICN-based IoT platform has been proposed [221]. NDN and MobilityFirst(MF) (MF is also one of the ICN architectures) cater to the need of both static and mobile devices. Comparision has been done between NDN architecture and MF architecture by the implementation of building management and bus management system scenarios. In bus management system buses are mobile devices and different sensors and actuators are static devices. From the observation, it was concluded that NDN outperforms MF when static devices are involved and MF outperforms NDN when mobile objects are involved.

3.3.4.6 In-network Computation in Edge Computing and Cloud Computing

In IoT infrastructure, the data collected from constraint oriented sensors are initially processed and then the refined data is transmitted to the requested host. This mechanism is called in-network computation in IoT. This helps to reduce the amount of data produced thus reducing the storage and high processing requirements. In-network computation also helps in simplifying the mobile node management, minimizing cached data, simplifying data routing and forwarding hence improving network life and battery life. And in-network computation paves the way for edge computing.

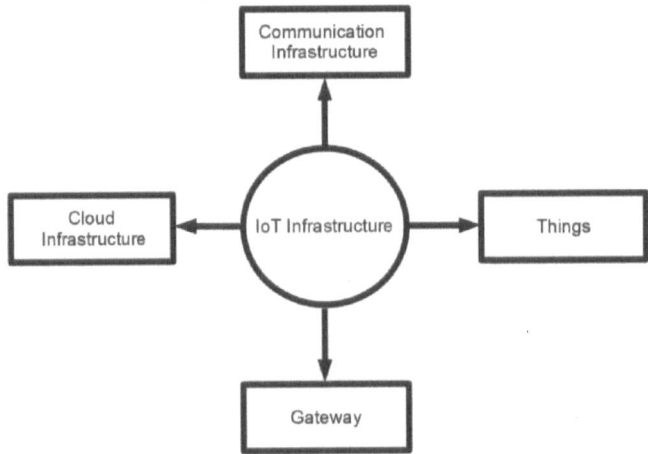

Figure 3.4: Requirements of IoT infrastructure

As shown in figure 3.4, cloud computing is the main driving force involved in the life cycle of IoT to process and manage IoT contents. But cloud computing creates a separation between the producer and the consumer of information. This results in increased delay and bandwidth requirement during transmission and reception of information to the central servers for cloud computing. Since the central servers are responsible for the processing of data and management of information so the intervention of central sever is necessary. Moreover, it may pose many privacy concerns during the reception and transmission of content. Considering these disadvantages, fog computing has been introduced to shift storage and computing capabilities towards the end node or edge node of the network. And since end node or edge nodes are involved fog computing is also known as edge computing [95]. Data needs to be cached before its processing in edge computing, and ICN-IoT allows to cache data naturally. Thus IoT devices can process the cached data in ICN-IoT as it allows caching with edge computing.

In ICN-IoT, it is encouraged to cache the data near the end consumers which increases the applicability of edge computing. Thus, in ICN-IoT caching, edge computing in-network computation plays a pivotal role. In IoT applications like virtual and augmented reality-based games which require real-time behavior with almost zero-delay can benefit from edge computing [307].

An alternative way to process and compute ICN-IoT data is by employing cloud computing [311]. The burden of processing can be shared by the cloud. Cloud also provides high storage and can also calculate analytics of any specific ICN-IoT application. As an example, the usage of electricity can be calculated and seen in any specific locality of the town. Hence, cloud-assisted ICN-IoT assists in designing systems that can perform complex calculations, provide huge storage capabilities and act as the backup in case of mobile devices.

3.4   Conclusion

The drawbacks of current Internet architecture and the significance of ICN in filling these gaps have been thoroughly discussed in this chapter. Moreover, the most prominent ICN architectures and future research challenges have been also discussed. The most prominent features of ICN like naming, forwarding and routing and in-network caching have been thoroughly covered. And considering the need to implement IoT in the ICN network the requirements and challenges to build a reliable and inter-operable communication network architecture for IoTs are presented. In the end, the role of edge computing and cloud computing in implementing the ICN-IoT network and future challenges have been also discussed.

Part II

Standards and Protocol

Chapter 4

Security in Cloud-Based IoT

Monjur Ahmed

Waikato Institute of Technology, New Zealand.

Nurul I Sarkar

Auckland University of Technology, New Zealand.

Cloud-based Internet-of-Things (IoT) is one of the emerging topics in IoT and Cloud Computing (CC). In addition to those found in CC and IoT, a Cloud-based IoT infrastructure inherits pros and cons of several other technologies and computing approaches. The distributed nature of a Cloud-based IoT infrastructure is prone to different threats and vulnerabilities related to technological and human-centric factors as well as strategic decisions in design and implementation of a Cloud-based IoT. This chapter explores Cloud-based IoT from security perspective. The technologies and design issues for Cloud-based IoT are explored in initial discussion. The security threats for a Cloud-based IoT infrastructure is addressed afterwards. Since CC plays a major role in Cloud-based IoT scenario, threats, security concerns, and vulnerabilities of CC are considered, which are then mapped into security threats and vulnerabilities for Cloud-based IoT. The security issues and threats are considered through three key factors, namely, technological, human, and governance. This chapter concludes by summarising findings and reflecting on key insights from the findings.

4.1   Introduction

Security is a major issue in computing. Since CC and IoT refer to specific computing approaches and technologies involving virtually all computing means and technologies, they inherit all computing security concerns, vulnerabilities and threats. In addition, CC and IoT come with their own additional security concerns. The added security issues result due to the way CC and IoT are designed and implemented. When the implementation of an IoT infrastructure embeds CC, the combined security concerns multiply. Security is thought to be the most significant barrier for CC adoption [93]. It can be assumed that security has similar significance for Cloud-based IoT due to CC being incorporated. A Cloud-based IoT is essentially an IoT infrastructure incorporating CC and IoT. In addition to its own security concerns, Cloud-based IoT inherits all security concerns found in CC and IoT.

Several taxonomies are developed for CC and IoT to realise how threats emerge in these computing approaches. No specific threat taxonomies for Cloud-based IoT exist to date; a Cloud-based IoT may be thought to inherit all security concerns that are inherent in CC and IoT itself. The existing taxonomies endeavour to note the factors behind threats for CC and IoT. These factors range from human-centric actions to technology-dependent loopholes [28].

Security concerns do not emerge only from technological loopholes or human-centric actions. The related rules, regulations and aspects of governance may also contribute the concerns of security and privacy for a computing infrastructure. This is no exception for Cloud-based IoT too. The technologies used, the infrastructural design and the regulatory environment within which a Cloud-based IoT operates, all these could be contributing factors for security concerns.

In this chapter, we discuss various security issues and challenges for Cloud-based IoT. The technologies, design aspects and challenges are discussed in the next section. The subsequent discussion addresses the security issues and threats for Cloud-based IoT. We also discuss the implementation aspects of Cloud-based IoT with a view to secure such infrastructure better. This chapter concludes by summarising findings and reflecting on key insights from the findings.

4.2   Motivation and Contribution

All kinds of computing that uses CC to any extent is prone to security concerns. Embedding CC in any computing architecture come with threefold security concerns - security concerns in traditional computing and computer networks, security concerns in the respective technology, and the additional security concerns of CC. Since CC embeds the use of remote resources that may be under someone else's management and ownership, and at the same time, the location of the Cloud servers may be situated in different geographic locations with different applicable laws and jurisdictions, the total security concerns for any architectural setting that incorporate CC thus are not only complex, but also come with direct, indirect and retrospective security concerns. As an emerging technology, IoT warrants investigation on its own. For CC embedded contexts of IoT – for example, Cloud-based IoT – and thus requires focus and exploration from a security perspective. The requirement to understand and realise security concerns, threats and vulnerabilities for Cloud-based IoT is the motivation for the discussion presented in this chapter.

The presented discussion helps to understand the technologies and design issues in Cloud-based IoT. This is further complemented by exploring security threats in CC, IoT and Cloud-based IoT. The crucial implementation aspects of a Cloud-based IoT infrastructure is helpful in deciding various performance of security related thresholds in designing and implementing such infrastructure. The regulatory context which encapsulates the technological setting of a Cloud-based architecture is also explored.

4.3   Research Method and Research Challenge

This chapter addresses the existing security aspects of Cloud-based IoT. As the focus of this chapter is to explore the security aspects and challenges as opposed to finding remedies for the security concerns and threats, our method is to collect and assimilate information through literature review. The findings from literature review are then blended with our own understanding and perception on Cloud-based IoT that is reflected and presented in this paper. Besides, the approach of IoT and any variant of IoT (e.g., Cloud-based IoT) are not mere technological deployment, they are also to be considered from strategic viewpoint when security is concerned. We have considered both the technological and strategic aspects of Cloud-based IoT.

Several contemporary technologies and topics overlap and there is a fine line among the operational aspects of these technologies and topics – this is a challenge to consider security aspects of Cloud-based IoT to ensure the focus is not deviated on security aspects of other overlapping technologies and topics. Examples of such technologies and topics are CC, Distributed Computing and Decentralised Computing. A proper balance is required to keep the focus specific to Cloud-based IoT without diverting much when discussing security aspects of other computing technologies and topics mentioned above.

4.4   Cloud-based IoT: Technologies and Design Issues

CC and IoT are the concepts of computing approaches rather than being computing technologies themselves. These computing approaches encapsulate all kinds of computing means to form infrastructure that are known as CC and IoT. Thus, it is imperative to first define CC and IoT before discussing the technologies and the design approaches of such technologies. In this section, we first define CC, IoT and Cloud-Based IoT. We then discuss the technologies used and design approaches taken to adopt, develop and implement CC, IoT and Cloud-Based IoT.

A number of definitions on CC and IoT exists. CC means using remote computing resources to satisfy computing needs of end-users, where the ownership and management of the remote computing resources belong to third parties known as Cloud Service Providers (CSP); and the end-users may not necessarily be aware of the particulars of their CSP [340]. Access to remote resources on CC are generally through public infrastructures, for example, the Internet [316]. A CC architecture can be geographically dispersed [238].

A formal definition of CC by NIST as mentioned in [84] defines CC to be a convenient way for on-demand computing minimal end-user effort, and with service providers' minimum interaction. A complete definition of CC is found in [27] which states, "CC is a conceptual computing approach that may encapsulate any other computing means and act as a wrapper for all kinds of computing practices. CC is the setting where hard (e.g., network infrastructure) and soft (e.g., data, software, processing) elements are remotely existent and access to these resources is on an ad hoc basis using public or private communication infrastructure, where the management and maintenance concerns of the Cloud infrastructure including the resources held within the infrastructure are most often beyond the end-users' scope". We agree that CC is rather a concept towards computing which can be used to adapt any computing approaches. Thus, an IoT infrastructure that uses remote resources through a CC infrastructure can be considered as Cloud-based IoT.

By definition, IoT means connecting anything and everything to the Internet, more specifically, the devices and gadgets people use in everyday life. IoT refers more to the consumer devices other than computers but those come with computing capability. It is vague and impossible to define what is not a computer in today's digital world. However, we can assume that the devices that have computing power embedded into them but are not used as computers may be thought of the 'things' within the context of IoT. Examples of these 'things' could be any device or component that has the capability to be connected to a network (though Internet is always used to denote a network for IoT), and can be monitored and/or controlled remotely. A few examples of 'things' (or devices) in IoT are smart watches, sensors, smart home appliances.

Given the above scenario to define what IoT is, it is apparent that wireless communications play an important role for an IoT infrastructure. This is important, as CC does not necessarily require wireless technologies to be qualified as CC, but IoT scenario mostly incorporate devices with wireless connectivity.

A question may arise as to why we need an IoT system. The simplest answer is, we want to have a network among tiny IoT 'things' or devices that come with limited power and capacity in terms of storage and computing power. The devices themselves thus can be part of an IoT infrastructure, but a sole device cannot act as the controlling node for an IoT infrastructure. This limitation is due to the devices being constrained in their computing power and storage capability, as mentioned above. This brings non-IoT 'things' or computing infrastructure (e.g., servers and their associated networking infrastructure) into picture; an IoT infrastructure may made up of millions of IoT 'things' and thus administering, monitoring and controlling such an infrastructure requires powerful computers. At this point, the option to explore is whether CC infrastructure can be used for such purpose triggering the introduction to Cloud-based IoT. Thus, when we consider Cloud-based IoT, we need to collectively consider the design aspects and technologies used in both IoT and CC.

The design of Cloud-based IoT incorporates the distributed nature of CC and the randomness of IoT devices. It is impossible and beyond anyone's control on how and when devices that connect to an IoT infrastructure will turn themselves on or off. An IoT infrastructure does not necessarily have to be made up of the devices that can be controlled; any foreign device can connect to an IoT infrastructure on an ad hoc basis. As virtually any technology can be used to deploy a CC or IoT infrastructure, we discuss the design issues that may emerge for an IoT context and infrastructure. IoT comes with its own challenges compared to traditional computer networks [89].

4.4.1 Design Issues

The key factors that may pose design challenges for a Cloud-based IoT are discussed below.

1. *Lack of Structuredness:* Considering the overall context of an IoT infrastructure, it must deal with ad hoc negotiation with nodes where the number of nodes or the amount of transaction may be well unpredictable. This brings the first challenge for an IoT to be designed optimally in terms of resource allocation and usage, depending on nature of the activities and goal of an IoT infrastructure.

2. *Device Limitations:* One aspect of the nodes of an IoT infrastructure is that the nodes are not designed for any specific IoT infrastructure. Rather, an IoT infrastructure needs to consider the devices that will be connected to the infrastructure, and the design and deployment of the infrastructure needs to be done accordingly. One of the major such considerations are the awareness of the capability of the devices that are aimed to be connected to the IoT infrastructure. The nodes or devices that connect to IoT are not

nodes of a computing network in any conventional meaning. The nodes stand on their own with distinct functionality and the connectivity and processing relating to be a node of an IoT infrastructure is an additional feature for any device. For example, a smart watch is not built to be a computer network node, but an IoT infrastructure needs to consider how it can provide functionality for a smart watch to be connected to it. The smart watch essentially needs to be designed as a smart watch, not as a computer node of a network, though it may come with Internet connectivity facility which makes it capable of being part of IoT. A smart watch design, for various reasons, must consider its main functionality that instead of being a powerful computer, for example, it needs to be small and lightweight enough for a human being to comfortably put it on. This example makes it evident that an IoT infrastructure needs to adopt itself to be suitable for the devices it aims to be connected to as nodes. The nodes of an IoT which are mostly devices/gadgets with limited computing power and limited battery power are one of the major factors that needs to be taken into account while designing an IoT infrastructure.

3. *Redundancy and Performance:* Redundancy (or additional provision of contingency resources) helps a computing infrastructure in several aspects, for example, making an infrastructure fail-safe or fail-proof improves availability and performance. For IoT devices, the concept of redundancy may need to carefully be considered while designing such infrastructure. As IoT devices are limited in terms of power and computing capabilities, and as the nodes are not traditional network nodes, redundancy at node level is simply impossible for an IoT infrastructure in terms of privacy and regulations, if not in terms of technology. On the other hand, even if it is possible for some contexts of IoT infrastructure to have redundant copies of data or nodes, it will subsequently introduce the performance overheads. The design of an IoT infrastructure is not a straightforward one. The challenge is to optimally design the infrastructure to accommodate various random end-user devices and not vice versa. The introduction of redundancy and the infrastructure controlling level requires careful examination of the performance overheads, latency, and location-based complexity that may result due to adding redundancy in an IoT infrastructure.

4. *Reliance on other Infrastructure:* The term "Cloud-based IoT" reveals a fact that an IoT may be well embedded or aggregated into other types of infrastructures (which is the CC in the case of Cloud-based IoT) for whatever is the reason. It may be assumed that such integration is either for the limitations of the IoT infrastructure or to boost IoT performance. In fact, an IoT infrastructure itself will be of poor performance and capability by itself considering the competitive edge of today's digital world, if IoT is not aided by other computing means. This is the reason Cloud-based IoT is in discussion and evolving as a popular approach for IoT infrastructure. However, this implies that IoT relies on other computing approaches or

infrastructures. This is one of the design challenges for IoT. It is imperative to consider other computing means when designing an IoT infrastructure. As mentioned, IoT may not boost to sustain without the help of other computing means like CC, Fog Computing for example.

5. *Privacy and Control over IoT Nodes:* IoT infrastructure may have to allow unknown devices to be connected. For a number of reasons, the infrastructure monitoring or controlling the IoT nodes may have little or almost no control on the nodes. While it is not possible to list numerous different aspects where this little or no control may bring operational and other challenges this is surely a design and implementation challenge for IoT. If a foreign device connects to an IoT infrastructure, low level of control to that node is either allowed or technologically possible. In such cases, the foreign node connecting to IoT may contain mechanisms or try to initiate activities that pose as unhealthy for the IoT infrastructure. While it cannot be controlled, the design and deployment of any IoT infrastructure need to address this issue.

6. *Device Location and Privacy:* In some contexts, the devices connected to IoT may be location independent, i.e., a device may connect from anywhere in the globe. How the device is connected to other networks or what is the computing context for the device being connected to IoT may remain anonymous or simply out of the controlling infrastructure's knowledge. The location of an IoT device is also related to the data privacy. For example, if an IoT infrastructure is breached through a node from a different geographic location, which laws and jurisdictions would apply to the incident (whether the IoT infrastructure's location or the device's location) is a concerning factor and design challenge for IoT. Unfortunately, there is no 'one solution fits all' approach when it comes to data privacy and related regulations. This remains as an open and all-time challenge for IoT design and deployment.

7. *Routing and Information Processing:* There are some contexts of IoT where information processing and routing becomes a crucial factor. One such example is Wireless Sensor Network (WSN). In an IoT context where various sensors are the nodes, the routing of information may be achieved using these nodes. These nodes come with limited computing and battery power as mentioned earlier. Thus, efficient design of such nodes involves minimal information processing and its resultant power usage. The controlling IoT infrastructure may not always be nearly located or it may not be possible to locate the controlling infrastructure near the nodes (e.g., if the nodes are deployed in a hard-to-reach location like undersea or in a battlefield). If an IoT relies on its nodes for information processing or routing purposes, the design of the IoT infrastructure needs to address this proactively.

8. *Wireless Technologies:* Though theoretically not a must, in practice the IoT nodes connect to its infrastructure mostly wirelessly. The interfacing of dif-

ferent wireless technologies embedded in IoT nodes are a design challenge. If the nodes are connecting to an IoT infrastructure, the holistic robustness (or lack thereof) will affect the infrastructure both in terms of performance and security. One of the aspects to be careful about IoT security is that, in addition to all threats, vulnerabilities and security concerns and different technologies used in IoT may come with their own loopholes. Wireless technology is such an example.

9. *CC Technologies:* CC is a booming technology that is gradually encapsulating all other computing approaches; IoT is no exception either. For a Cloud-based IoT, the specific management and operational aspects of Cloud infrastructure as well as technologies used in CC are factors to consider. For Cloud-based IoT, the issue and challenges for CC are additional challenges not only from performance and compatibility perspectives, but also from security point-of-view. The security related concerns for Cloud-based IoT are discussed in later in this chapter. CC may use different Operating Systems (OS), virtualisation technologies, and the location of CSP data centres may be geographically dispersed. For IoT contexts where real time processing is a requirement, the dependency on Cloud infrastructure may pose as a design challenge.

10. *Technology Compatibility:* IoT is heterogenous in terms of technology. Different technologies are used in different devices. As it may not be possible to limit or control the number of type of devices for some IoT context, device and/or technologies used in the devices and their compatibility (hard or soft compatibility) can be a design challenge for IoT infrastructure.

11. *Distributed Architecture:* CC is distributed in nature. Subsequently, Cloud-based IoT is bound to be distributed in nature. The distributed nature of CC and IoT brings challenges in terms of security and privacy. Distributed resources are complex to manage and prone to vulnerabilities and threats. Efficient provisioning of CC and Cloud-based IoT has no provision but be decentralised and distributed. This inherently comes with the challenge of managing distributed resources and maximising security-integrity of these distributed resources.

The design challenges for IoT (or any other computing approach) do not stand only within the technological aspects, but also human aspects too. One of the major concerns or design challenges is the security concerns. Security concerns are vast and thus we address it in a different section that follows.

4.5   Cloud-Based IoT: Security Threats

Since Cloud-based IoT uses Cloud infrastructure, the security threats and concerns for CC apply to it, in addition to its own security concerns. For this

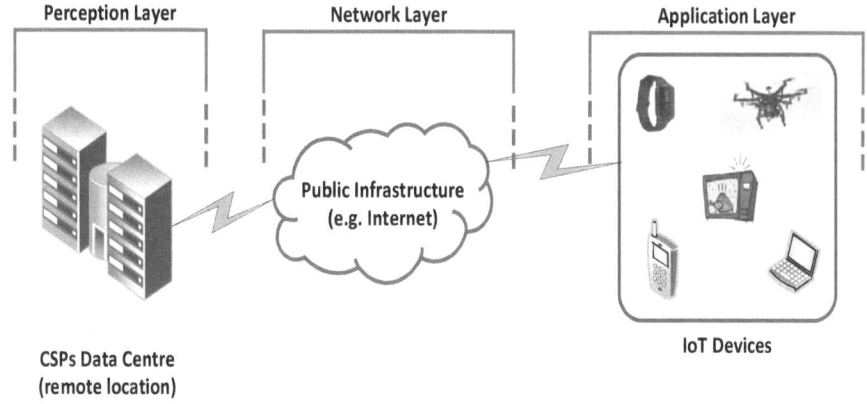

Figure 4.1: A Typical Cloud-based IoT Scenario

reason, we discuss security threats and concerns of Cloud Computing and IoT in a distinct manner which collectively forms the security concerns and threats for a Cloud-based IoT. Several taxonomies for both CC and IoT exist to realise the facets from which security threats may emerge. We consider few taxonomies as foundation of our discussion in this section. We discuss the major areas of security concern for CC instead of listing specific threats. Instead of listing the non-exhaustive list of all specific threats (which is a massive list and is redundant to list here), we point to the generic areas where threat vector may exist for a CC infrastructure.

From a contextual view, IoT is perceived to have three layers, namely, application layer, network layer and perception layer [242]. The application layer consists of the end-user IoT devices, and the perception layer is the infrastructure (e.g., servers, computer network and other terminals to receive and use data from IoT devices) which is the base that provides or collects (or both) data to and from IoT devices, as well as controls, manages and administers the overall IoT infrastructure. The network layer is the portion through which the data traverses between the application layer and the perception layer. In a Cloud-based IoT scenario, the perception layer exists in a Cloud infrastructure to be more precise, in Cloud servers of a CSP's data centre(s). A typical Cloud-based IoT scenario is illustrated in Figure 4.1.

Referring to Figure 4.1, a Cloud-based IoT essentially incorporates the IoT infrastructure at Application Layer level, the public network infrastructure (e.g., Internet) at Network Layer level, and CC infrastructure at Perception Layer level.

As illustrated, the Application Layer of a Cloud-based IoT is the operational layer consisting of the end-users' devices or gadgets. The Network Layer is normally a public communication infrastructure (e.g., the Internet) through which data is transmitted back and forth between remote Cloud servers and the IoT devices of the Application Layer. The remote Cloud servers reside at the Perception Layer, where collected data (from Application Layer and transmitted through Network Layer) is stored and processed. Given the above scenario, a Cloud-based IoT is potentially exposed to and associated with all threats and vulnerabilities found in IoT, as well as those found in computing network infrastructures and CC infrastructures.

4.5.1   Cloud Security Threats

Security is probably the most sensitive and biggest concern for CC. A number of taxonomies exist for CC threats. The proponents of CC threat taxonomies look into different perspective when looked at CC threats. We take a threat taxonomy found in [27] (by one of the authors of this chapter) as a foundation of our discussion in this section, which is a genre-based threat taxonomy for CC. The threats for CC emerge from either technological or human-centric factors or both. Internet-based vulnerabilities prevail in any kind of security concerns, due to the fact stated in [146] that Internet is the primary means to access Cloud resources.

The technological threat factors for CC arise from both hardware and software used in a CC architecture. The use of local platform and those used in Cloud servers are a source of threats in that they may contain security loopholes. If the IoT controlling infrastructure is Cloud-based, the OS used in the CC infrastructure is thus of concern. The same applies to various network protocols used within a CC setting. If a network protocol used in CC come with security threats, this will retrospectively affect an IoT if housed in that CC infrastructure. The same applies to any software tool used in a CC. The existent loopholes in software tools used in CC have the potential to affect the integrity of an IoT deployed on that CC.

Virtualisation is a core technology behind CC's massive popularity. When considering CC, all aspects including the security concerns of virtualisation should be considered. Virtualisation is associated with several software-based security threats [304], for example Denial-of-Service (DoS) and hypervisor exploits. Since virtualisation is founded on resource sharing, the shared hardware environment is a potential security threat point for CC and anything built upon CC, let alone Cloud-based IoT. Apart from the generic security concerns for virtualisation, the hypervisor used in CC may come with its own specific threat vector.

Considering Cloud-based IoT, web services are used in CC and eventually in IoT as an unavoidable part. Webservices come with potential security loopholes, where Hyper Text Transfer Protocol (HTTP) vulnerabilities could be a good example [273]. The security mechanisms used for communications within a CC or IoT infrastructure may themselves be the source of breaches. Cryptographic mechanisms used for data encryption may have their own weaknesses [127] leading to opening security loopholes.

Mobile computing and portable devices are inevitable part of IoT. This is true for CC contexts as well, for example, the scenario of Bring Your Own Device (BYOD). Mobile computing is becoming common means for general as well as distributed computing [97]. Security vulnerabilities in mobile technologies are thus introducing vulnerabilities in CC and IoT, and needless to say in Cloud-based IoT. Mobile computing introduces application-based and network-based threats [131, 204, 367] in computing infrastructures.

Failing or malfunctioning hardware may introduce vulnerabilities within a CC infrastructure which subsequently will affect any IoT associated with that CC infrastructure. Besides, maintenance service or incident response scenario may open up opportunity for an insider or outside person to access resources that they are not normally entitled to, leading to opening up opportunity for malicious activities. Additionally, outsourced Cloud services may lead to insider attack through on-premise or remote access to Cloud resources [122]. Considering mobile phones as nodes of a distributed computing environment like CC or IoT (and, of course, Cloud-based IoT), it is important to note that mobile phone users have poor control over their phone and the apps they use [131] as smartphone apps tend to access more data and have more control than required for their functioning.

Human factor-centric threats prevail over technological factor-centric threats in that the former is mostly unpredictable. Human factor-centric threats are terms as soft threats, and genre-based example of such threats are trust, compliance, regulations, competence and specialization, Service Level Agreement (SLA) mis-interpretation and Social context including social engineering [27]. It is important for CC (and hence Cloud-based IoT) users to be competent and compliant. Social engineering is a constant threat for Cloud environment [202]. Human-level trust and social engineering are social factors that cannot be countermeasures in systems design and implementation; rather a soft skills requirement for the users would be more beneficial this is quite a challenging threat area for CC and IoT.

Location transparency is a security concern for CC. For Cloud-based IoT, it poses twofold security concerns, both for the CSP and for the end-users. An IoT device from unknown location may be used to compromise a CC infrastructure. On the other hand, an unknown location of the CSP may pose privacy and security concerns for end-users of a Cloud-based IoT.

CC (and this Cloud-based IoT) is complex in terms of architecture. This complexity may affect the expectations of stakeholders when it comes to the monitoring, enforcement and expectation from SLAs. Arguments exist that SLA provisioning required careful consideration [83]. The confusion or unrealistic expectation from SLA may result in trust issues in CC settings and thus trust management becomes a key issue [143]. Cloud users (hence, Cloud-based IoT users) trust (or expected to trust) CSPs with their sensitive confidential data [315]. This trust must not be a blind trust; it should rather be based on the overall integrity of a CSP. It is a concern on how CSPs apply compliance standards in any specific geographic location [337]. Based on the above, it can be assumed that regulations governing CSPs in different regions may also be sources of security concerns for any services that are based on or related to CC.

The above discussion points to the major areas of concerns related to CC security. Developing a specific list of CC threats depends both on soft and technological factors. The list thus will always be a time and location dependent non-exhaustive long list.

4.5.2   IoT and Cloud-based IoT Security Threats

Cloud-based IoT brings all the security challenges of CC that were discussed earlier. In addition to that, we look at IoT specific security concerns in this section. IoT is defined as a network of everyday things connected among themselves through Internet [38]. Considering Cloud-based IoT apart from CC security concerns any threats or security concerns about the Internet applies. On top of inheriting threats from Internet and CC, Cloud-based IoT may also be exposed to technology specific threats. For example, an IoT infrastructure that deploys Radio-frequency Identification (RFID) technology is susceptible to jamming attack [90] which is an RFID-specific attack. As a result, the Cloud-based IoT can be considered to be prone to countless threats and vulnerabilities. While a Cloud-based IoT brings its own benefits, it is probably a very significant quest to decide whether Cloud-based IoT will actually be beneficial in the long run. We are not against implementing Cloud-based IoT, but we are skeptical about IoT integrity including privacy and security if implemented using public Cloud.

Managing access control, authentication and authorisation is mentioned as some of the sources of security concerns for IoT [38]. This is mostly due to limited capability of the IoT devices and an inherent design challenge and are potential points for threats vectors to emerge. IoT is perceived to have a layered architecture [242] and each layer comes with its own security threats. IoT security threats are those found in CC, Internet and computing in general. Trust and privacy issues raise security concern in IoT [324]. The attacks for IoT mentioned in existing literature are existent attacks that were not unknown. Examples of such attacks are malware attacks, Denial-of-Service (DoS) and Distributed DoS (DDoS), identity theft, sybil attack, eavesdropping, and spoofing attack to name a few. An architecture-based list of attacks is presented in [90] which lists IoT attacks based on its four architectural layers, namely, application layer, middleware layer, network layer and perception layer. We do not list or discuss all the mentioned attacks to avoid repetition and redundancy, as they are as mentioned earlier known attacks for any computing settings. However, one attack mentioned in [90] can be considered as IoT specific. This is sleep deprivation attack done on the IoT devices to make it unable to go to 'sleep' mode in order to have the devices' power consumed faster.

Instead of producing a long list of IoT security threats (and repeating discussion thereof), it is enough to say that all known security threats in computing and CC apply to a Cloud-based IoT. To sum up, a Cloud-based IoT inherits security concerns, threats and vulnerabilities from the Internet, CC and all technologies that are used within an IoT context. IoT itself is not a technology, rather a concept. The concept of IoT is actualised using computing and other technologies. This makes is harder and challenging to generalise threats and security concerns for IoT. While all existing concerns from computing approaches including the Internet and CC should be taken into account, IoT infrastructure remains vulnerable to further security concerns depending on what other technologies are used in an IoT infrastructure. As mentioned earlier, examples of such technologies could be RFID, various sensors and their related technologies that may bring their own vulnerabilities to introduce within an IoT context. The design challenges for IoT is discussed earlier and it need further explore to result the vulnerabilities for a Cloud-based IoT.

4.6   Implementation aspects of Cloud-based IoT

The goal of this chapter is not to discuss or present ways to deploy an IoT or Cloud-based IoT infrastructure. Yet, the deployment of any computing infrastructure requires strategic insight from a security perspective. Cloud-based IoT is no exception. While choosing technology may have impact on technological security vectors or vulnerabilities, the soft factors (or strategic factors) are important to avoid any retrospective security concerns that may emerge even without the knowledge of the people involved in designing and implementing a Cloud-based IoT infrastructure. From security viewpoint, we discuss the major strategic implementation factors for a Cloud-based IoT Distributed Architecture, Privacy, Environment, and Regulation & Compliance. It is important to note that, since Cloud-based IoT inherits all security concerns of CC and possibly those inherent in all other computing means, the list or categories of security concerns, threats and vulnerabilities will remain ever non-exhaustive. This is because as mentioned earlier CC and IoT are not computing technologies; they are rather conceptual computing approaches that may encapsulate all other computing means and other associated technologies.

1. *Distributed architecture:* CC and IoT come with distributed architecture. By nature, CC and IoT may have their resources distributed and the operating boundary infrastructure may be vague. This makes the management and maintenance of such infrastructure a very complex task. The distribution of resources introduces numerous attack vectors within a CC or IoT context. If the cause is distributed in a computing architecture, the effect will subsequently be distributed [27]. Thus, distributed security is essential for a distributed architecture. For CC and Cloud-based IoT, a distributed security mechanism or model is required. Due to the complex and distributed nature of CC and CC based IoT, the security measures should

also be distributed. The distributed security mechanism in CC also needs to be decentralised [27]. Some research endeavours propose distributed and decentralised security models for distributed computing infrastructure. Implementation of distributed and decentralised security is not practiced enough to date. One example of distributed and decentralised security model is proposed in [27]. Other examples of distributed or partially distributed security model or mechanism for CC are found in [84] and [231]. When designing and deploying a Cloud-based IoT infrastructure, the distributed and decentralised aspects of security are to be taken into serious consideration.

2. *Privacy:* Privacy is another issue and strategic threat vector for Cloud-based IoT that requires full attention while designing or implementing an IoT infrastructure. Both the technological and human aspects to privacy related threat vectors are design and implementation challenges for any distributed computing infrastructure. This becomes more challenging when the infrastructure spans over different geographic locations. Though virtualisation is a revolutionary technology behind massive popularity of CC, it poses threats due to its nature of shared (among different customers) hardware and software environment within a Cloud server. On the other hand, the definition and perception of privacy may differ from one geographic location to another either due to local law or cultural difference, or for any other reason. This needs to be taken into account in deploying a distributed architecture like Cloud-based IoT. The major challenge in this regard is that there is no recommended solution as the solution or best practice to ensure integrity of privacy entirely depends on each case/context individually.

3. *Environment:* IoT (whether Cloud-based or not) heavily depends on wireless technologies. An example of such dependency of IoT on wireless technologies is found in the discussions in [264]. Some IoT infrastructure may be deployed within the proximity of other such infrastructures and the devices and traffic of a certain IoT may be exposed to other surrounding or overlapping infrastructures operating using wireless technologies. This is an implementation factor for any kind of IoT infrastructure regardless, including Cloud-based IoT.

4. *Regulation & Compliance:* Privacy related factors and those associated with regulations & compliance are interrelated. When it comes to the regulation and compliance, one challenge for CC (and subsequently, Cloud-based IoT) is the geographic dispersion of a CSP's Cloud infrastructure. A CSP may have its office in one country and data centres in different countries. It is important to ascertain where the data will be stored and through which regions the data will be in transit for a Cloud-based IoT infrastructure. This will further need to complement understanding which region's regulations and jurisdictions are applicable in case of a data breach. This needs to be very specifically and clearly defined in the SLA. This is probably the

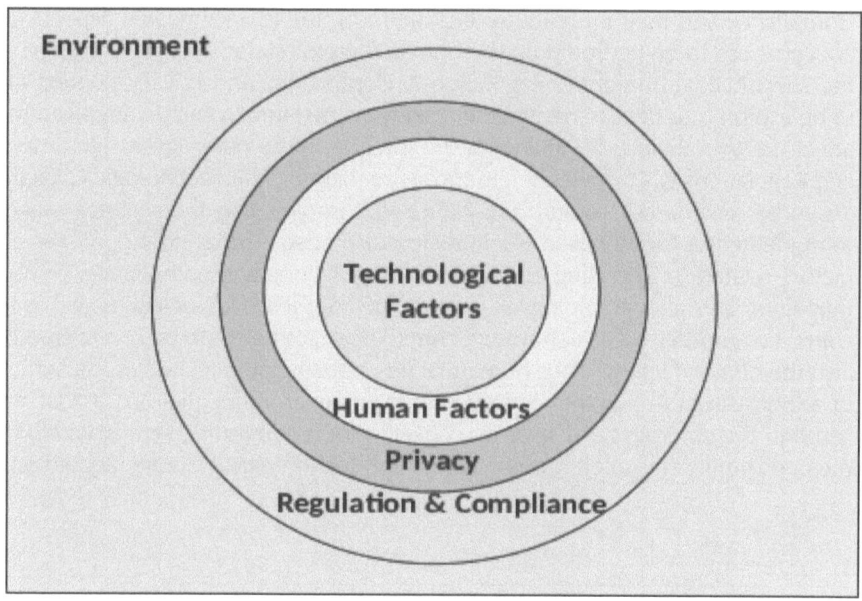

Figure 4.2: Framework for Cloud-Based IoT Security Analysis

most important aspect while designing and deploying a Cloud-based IoT infrastructure.

One of the main implementation issues of Cloud-based IoT is that the topic is broad with no proposed and uniform architecture [239]. While data security and privacy are important factors to consider for the implementation of Cloud-based IoT [267], the actual implementation requires strategic analysis from both technological and regulatory perspectives. A layered strategic approach to design and implementation may provide an integrated and structured approach. In this chapter, we propose a strategic layered framework of security analysis in designing and implementing a Cloud-based IoT. This is illustrated in Figure 4.2.

As illustrated in Figure 4.2, the security threats, concerns and vulnerabilities in designing and implementing Cloud-based IoT need to address the four major aspects in a layered manner, namely, Technological factors, Human Factors, Privacy, and Regulation & Compliance. These factors are to be taken into account in addition to and within the Environmental context where the Cloud-based IoT infrastructure will be implemented. Any infrastructure development needs to first consider the environment and the existing other computing infrastructures (e.g., wireless infrastructure) that may pose a threat or security concerns to the IoT network to be implemented. Regulations & Compliance factors of the locality of the IoT to be implemented, of the CSP and its data centres to be used, and of the locations from the end-user devices are likely to access the network need to be considered. Within the boundary of identified Regulations &

Compliance and their integrity (or lack thereof), the Privacy related aspects and concerns are to be figured out. The human factors related security concerns for the specific Environmental, Regulation & Compliance and Privacy context also to be explored to determine whether and how the Human Factors involved may affect the overall security including the integrity of its outer layers (i.e. Privacy and Regulations & Compliance). In the same manner, the Technological, Factors are to be scrutinised. To sum up, Figure 4.2 implies that the security analysis to implement a Cloud-based IoT infrastructure needs to consider concerns and factors related to operating environment, regulations and compliance, privacy, human models and technological factors. Each of these factors stands within the context of another – as illustrated; technological factors are to be considered but also the human factors, since human factors influence the technological settings of an infrastructure. In the same way, privacy is an influential factor that may result in the emergence of security concerns for a computing setting which has the opportunity to exploit human factors and technological factors, and so on.

4.7 Concluding Remarks

The practice of secured computing within the context of CC and IoT is inadequate to date. The state of security in CC is doubtfull to be confusing by some researchers [129]. IoT lacks well investigated security and privacy guidelines [19]. This leads to the realisation that the state of security for Cloud-based IoT has a long way to go. The current provisioning and deployment of Cloud-based IoT thus needs to take a careful approach and an architecture wide exploration of CC, IoT and all relevant technologies that would be used in any specific IoT context. In addition, while the design challenges may not be avoided in deploying Cloud-based IoT, the security concerns and possible vulnerabilities embedded in implementing IoT infrastructures must be taken into serious consideration.

CC, IoT and Cloud-based IoT, all these conceptual computing approaches, are prone to both technological and human factor related threats. Technological factors are mostly structured and quantifiable while the human-factors related threats are mostly unstructured and mostly qualitative. Technological threats are given most focus and importance, but the strategic and human-centric threat vectors are not investigated enough for distributed and geographically dispersed infrastructure that are inherent with any CC-based system. The human-centric factors and strategic aspects of designing and deploying Cloud-based IoT thus requires more attention.

Cloud-based IoT inherits all advantages as well as challenges of both CC and IoT. As revealed throughout discussion in this chapter, the security concerns for a Cloud-based IoT infrastructure come from technological, human-centric and strategic decision-making related factors. Technologies used within the context are of concern, so are the CSPs involved and the strategic design decisions employed in designing and implementing a distributed computing infrastructure like Cloud-based IoT. The inheritance of complex resource distribution from CC multiplies vulnerabilities for a Cloud-based IoT compared to a traditional IoT scenario. The discussion presented in this chapter reveals that the strategic and human-centric factors may be more detrimental for Cloud-based IoT security compared to those that emerge from technological factors.

Chapter 5

Cloud Enabled Body Area Network

Anupam Pattanayak

Indian Institute of Information Technology Guwahati

Subhasish Dhal

Indian Institute of Information Technology Guwahati

5.1 Abstract

There is a proverb that says *prevention is better than cure*. Doctors usually suggest several measures such as regular medical check up, clinical tests, physical exercises, healthy diet, etc., to prevent diseases. However, with the enormous advancements in computing and healthcare technologies, it is of natural demand that a person, facing any sudden medical emergency, be reported instantly and automatically to all the concerned parties such as the patient herself/himself, medical service provider (MSP), doctors, etc. This is made possible by use of a newly emerged technology called Body Area Network (BAN). This technology consists of different sensor nodes implanted in the human body to read physiological health information (PHI) parameters such as blood sugar, blood pressures, body temperature, etc. If any PHI reading is beyond the normal range of the corresponding PHI, the medical event is reported instantly. However, the minute monitoring and detail analysis of PHI data 24×7 requires huge computational resources, which is beyond the reach of BAN. An appropriate Cloud environment can be an aid to overcome this computational limitation. In this mechanism, a medical user (MU) empowered with BAN is connected with a Cloud (also can be termed as health Cloud) environment. BAN attached to the medical user continuously sends the physiological readings to the Cloud, which in turn does all the tasks of data analysis and reports to all the concerned parties, whenever there is any medical abnormality found by the analysis. The MSPs are, on the other hand, registered with the same Cloud to provide services to the MU remotely such as monitoring the MU's physiological status when hospitalization is not necessary for continuing the treatment. MSPs also can use Cloud aided BAN technology for the MUs admitted in their own healthcare facility. In case of any medical abnormality, the selection of appropriate MSP, depending on the nature of ailment, is also done by Cloud automatically. However, one of the essential requirements in this technology is both MU and MSP have to share several sensitive information such as real identity, physical location, physiological parameters, infrastructural status, etc., with the Cloud. These can become threats in several forms for the users of this nice and useful technology. In this chapter, the several kinds of threats against the security and privacy of the users of Cloud aided BAN are discussed. Also, it briefly enlightens the research attempts that tried to thwart these threats and enable this very useful technology for use in a smart society.

5.2   Introduction

The term *Computer Network* (or simply, *Network*) is a common word in the field of Computer Science and Electronics. Network is the wired or wireless interconnection among the computing devices and switching elements that enables communication among the devices. It is also the backbone of today's revolution in information and communication technology. Wireless Sensor Network (WSN) is a kind of network, where sensors are the communicating devices. Sensors are the tiny electronic devices that can sense some physical characteristics such as temperature, air moisture, wind speed, wind direction, physical movements, etc. A block diagram of a sensor node is shown in Figure 5.1.

Different types of sensors used in medical fields can be classified into three categories: physiological sensors, bio-kinetic sensors, and ambient sensors. Physiological sensors measure blood pressure, blood glucose, body temperature,

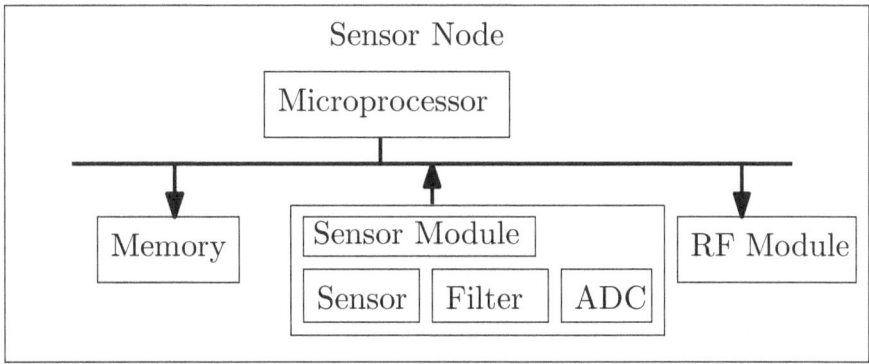

Figure 5.1: Block Diagram of a typical Sensor Node

blood oxygen, etc. Bio-kinetic sensors measure human movement in terms of acceleration and angular rate of rotation while in motion or movement. Ambient sensors are those which can measure environmental parameters such as temperature, humidity, light, and sound pressure level. Over the past decades, several bio-sensors (or body sensors) have been developed that can read different physiological parameters such as body temperature, blood glucose, blood pressures, etc. These bio-sensors in a human body form a special kind of network, known as body area network (BAN) or wireless body area network (WBAN). A sink node or gateway node is placed in the human body to accumulate the physiological readings. A hand-held device such as smart phone, or a smart wrist watch, connected to the Internet forwards these physiological data to servers in hospitals or such service providers for analysis. However, all the hospitals may not be well equipped for this continuous data analysis because it involves investment in computing resources, infrastructure, and skilled personnel. Therefore, Cloud is an alternative for this continuous physiological condition monitoring without fail. This health Cloud can be public Cloud or private Cloud. To restrict the scope of our discussion, we assume the health Cloud as public Cloud. If there is any abnormality in the physiological reading of a user, then that is reported instantly to all concerned parties so that medical treatment can be started as early as possible.

As society is evolving towards smart society, several countries are taking up projects to make some of its cities smart cities. Hospitals are also using more and more smart technology-enabled services to cater to its patients. This can be better explained with some examples.

I. Suppose a patient is hospitalized in a critical care unit with severe medical problems, and has been stabilized after few days of medical treatment. Therefore, the same critical care unit can be vacated and allocated to another severe patient, and the patient in stable condition can be shifted to general bed for further observation and treatment. However, patients in the general beds need to be monitored with proper care as the patient is still not fully recovered. Since the general beds do not have facilities like critical care units, BAN can help to monitor this patient continuously. Thus the critical care unit can be utilized for multiple patients without any lack of care to any patient.

II. The aged population is increasing day by day. There are many aged persons, who are otherwise medically fit, but can suffer a lot if they fall on the floor by losing physical balance. This sudden fall, if not detected early, can delay the proper treatment and can become very fatal. If the elderly people are powered with BAN, then medical care to them can be much more effective.

III. Many people usually suffer cardiac problem or cerebral attack during their sleep, where they succumb to death most of the times without any medical treatment. This can be prevented at large if a person is equipped with BAN.

IV. Similarly, parents feel helpless and become worried because their children cannot express the nature of their physical uneasiness. The anxiety of parents and sufferings of children can be reduced if the reason for their physical uneasiness can be instantly detected by cloud enabled BAN.

V. Next, consider a situation where a person is met with a severe road accident with nobody around to notify the family members or the doctors. And the physical condition of the person is so critical that, if the person does not get quick medical treatment, the result may be catastrophic. If the person is equipped with BAN, and the BAN is connected to a Cloud, the medical emergency of the person can be immediately analysed and detected by the Cloud, which in turn can notify the nearest trauma care service provider for sending ambulance to the accident spot without any delay. This can largely speed up the treatment of the patient and thus life of the patient can become out of danger.

VI. In another example, consider the case that many road accidents occur in the late night or early morning due to the fact that the driver falls asleep at the wheel. As a result, the car may collide with some other vehicle or objects like a tree. Even if the driver closes his eyes for a fractions of a second, there may be a catastrophic mishap. This type of accident can be reduced if the driver is equipped with BAN having a sensor for monitoring sleep or awake status.

Thus there are several examples where BAN can immensely help in proper and efficient treatment. These examples can serve as the motivation for a smart society to adopt cloud-enabled BAN for detecting most of the medical problems instantly. So far, we have discussed only the application of BAN for continuous monitoring of health data. However, BAN can also be used in so many other applications. These include sports training, military applications, medical automation system, human computer interaction systems such as gesture detection system and emotion recognition system, information security and forensic systems such as deception detection system and authentication system, and education systems such as smart tutoring system and teaching assistant system [278].

Although Cloud based BAN is of great help in a medical emergency, the involvement of a third party, i.e., Cloud, can become a security and privacy risk for the users. The medical behaviour of a user can be easily eavesdropped and exploited. The research community considered this as a very important aspect to deploy Cloud based BAN in healthcare. In the next few sections, we will gradually highlight these issues and their countermeasures available in the literature.

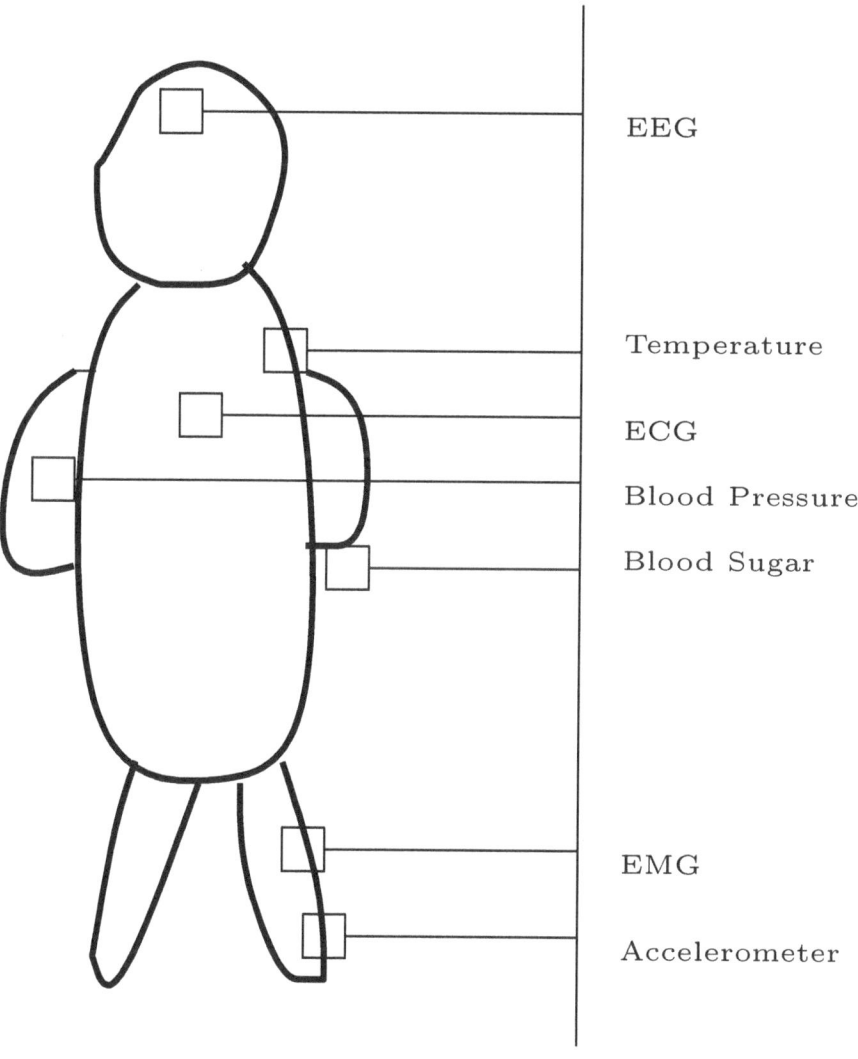

EEG

Temperature

ECG

Blood Pressure

Blood Sugar

EMG

Accelerometer

Figure 5.2: A human body with some sensor nodes

5.3   Bio-Sensor Nodes

A bio-sensor node has two parts: physiological signal sensor and the radio platform. Function of body sensor is to read analogue signals corresponding to humans' physiological activity. This analogue signal is digitized and then forwarded by the radio transceiver to the sink node attached to the human body. Some sensors are required to be implanted inside the human body, and some sensors are usually placed on the surface of the human body. Figure 5.2 shows a diagram of a human body with some bio-sensors. To name some sensor devices for BANs, we can mention the following: blood glucose sensor, blood pressure sensor, CO_2 gas sensor, accelerometer sensor, ECG sensor, EEG sensor, EMG sensor, gyroscope sensor, pulse oximetry sensor. Level of glucose in blood, also known as blood sugar, is an important parameter to get the physical status of any person. Both kinds of sensor, a non-invasive sensor that can be put onto the body surface and implantable sensor that are implanted in the body, are available for monitoring the blood sugar level. Non-invasive sugar monitoring is made possible by using infra-red technology and optical sensing. Blood pressure measures the force of blood flow on the blood vessel wall. Two kinds of blood pressure measurements are taken: systolic pressure, and diastolic pressure. There are sensors to measure both the pressures. CO_2 gas sensor reads carbon dioxide levels in blood and also reads concentration level of oxygen during respiration cycle. Pulse Oximetry sensor measures the SpO_2 level in blood by photoplethysmograph (PPG) signal. Heart muscles continuous contraction and expansion produce specific signals that are captured by Electrocardiogram (ECG). ECG is used to investigate heart condition. Body muscles produce specific signals during contraction or rest. Electromyography (EMG) captures this signal emitting from body muscles. As nerves control the muscles responses, study of nerves is done by EMG. Electroencephalography (EEG) monitors the electrical activity within the brain by placing small electrodes on the humans' scalp at different locations. Different sensors are available for ECG, EMG, and EEG. The activity and the motion of a person can be monitored by accelerometer and gyroscope. Accelerometer sensor measures acceleration. Gyroscope sensor reads angular velocity. Table 5.1 summarizes a list of different types of sensors available commercially.

Conventional operating systems such as Linux, Windows cannot be adopted by sensor nodes in BAN because the sensor nodes are resource constrained. Most of the sensor nodes use Tiny OS as the underlying operating system. Apart from Tiny OS, other alternative choices of operating system for sensor nodes are Contiki, Mbed, nano-RK, and FreeRTOS [284].

Table 5.1: Common Bio-sensors

Sensor	Sensory Measurement Description
Accelerometer sensor	Acceleration caused by user's movement
Blood pressure sensor	Systolic and diastolic blood pressures
CO_2 gas sensor	CO_2 level in blood
ECG sensor	Electrical activity of heart
EEG sensor	Electrical activity of brain
EMG sensor	Electrical activity of skeletal muscles
Galvanic skin response senosr	Electrical characteristics variation of skin
Glucometer	Blood sugar concentration
Gyroscope sensor	Change in orientation due to user's movement
Heart rate sensor	Heart contraction count per minute
Humidity sensor	Humidity in subject's surroundings
Magnetometer	Specifies direction of user
Microphone	Accoustic sounds created by user in awake or sleep state
Oximeter	Oxygen saturated hemoglobin
Pedometer	Step count of user in motion
Plethysmogram	Change in pulsatile blood flow
Respiration rate sensor	Chest rise count per minute
Spirometer	Respiratory flow rate
Strain sensor	Strain in different parts of body
Thermometer	Body temperature

5.4 Body Area Network

Body area network is one of the most active interdisciplinary research areas of Electronics and Computer Science. Several top class survey papers exist in the literature such as [92], [213], [281], [155], [209], [243], [191]. One of the earliest survey works on BAN was done in [92], and they have discussed design of MAC layer, physical layer, and radio technologies used in BAN. They have compared existing body sensor nodes, their operating systems, radio technologies used, data rate, outdoor communication range. They also studied several BAN projects taken up for remotely monitoring health or fitness status. Another excellent survey work has been performed in [213]. They have nicely studied the architecture of BAN, data rate of different types of sensors, energy consumption, positioning of BAN, characteristics of three layers: physical, MAC and network. They also mentioned about IEEE 802.15.6, BAN standard, which was then in the formative stage. They also touched on the security issues in BAN. We can mention here some other IEEE 802 standards, which are given in the Table 5.2.

Table 5.2: IEEE 802 Standards

IEEE Standard	Purpose	Status
IEEE 802.1	Higher Layer LAN Protocols Working Group	Active
IEEE 802.2	Link Layer Control (LLC)	Disbanded
IEEE 802.3	Ethernet	Active
IEEE 802.4	Token Bus	Disbanded
IEEE 802.5	Token Ring	Disbanded
IEEE 802.11	Wireless LAN	Active
IEEE 802.15	Wireless Personal Area Network (WPAN)	Active
IEEE 802.15.1	Bluetooth	Disbanded
IEEE 802.15.3	High-rate WPAN such as Ultra Wide Band (UWB)	Active
IEEE 802.15.4	Low-rate WPAN such as ZigBee	Active
IEEE 802.15.6	BAN	Active

The survey in [281] is another excellent work. They have discussed the IEEE 802.15.6 standard for BAN in detail. They have also discussed the communication architecture of BAN, layers of BAN, routing protocols in BAN and its classification and challenges, channel models, interference issues, antenna design, security and privacy issues. They highlighted the open problems in BAN. The research in [243] studied BAN from the angle of Wireless Sensor Networks and highlighted BAN research challenges.

As sensors and base stations are resource-constrained, Cloud is used to receive the sensory data from BAN and performs resource-intensive data processing of long-term storage of mammoth data. Furthermore, Cloud provides access to shared resources to BAN-based applications in a pervasive manner. Also, Cloud-based BAN offers the facility for remote update/upgrade of software in BAN. This makes maintenance of BAN more quick and cost-effective [278]. Review of Cloud-assisted BAN and its challenges have been discussed in [133]. Integration of BANs with Cloud infrastructure raises the following research issues apart from security and privacy[133]:

I. Interfacing Cloud with BAN

II. Sensor stream management

III. Massive scale and real-time processing

IV. Advanced off-line data analysis.

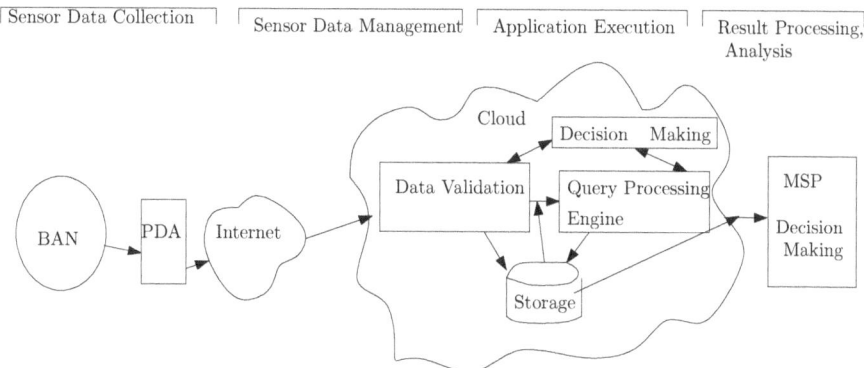

Figure 5.3: A Reference Architecture to integrate Cloud with BAN

The reference architecture for integrating BAN and Cloud is shown in Figure 5.3.

To mention a few BAN based research projects, we can name *CodeBlue*, *AID-N*, *SMART*, and *CareNet* [92], and *MediNet* project launched by Microsoft in Caribbean countries for remote monitoring of diabetes and cardiovascular diseases [225].

5.4.1 Communication Architecture

The communication in BAN is wireless. The three-tier communication architecture of Cloud-enabled BAN has been depicted in the Figure 5.4. The innermost communication is the tier-1 communication. Here, sensors are scattered throughout the body and send the PHI readings to the personal server in BAN, while the person may be in any posture like working, running, walking, sitting, or sleeping. The tier-2 communication is between BANs. One BAN can communicate with the other BAN. Also, tier-2 facilitates connection of BAN with other kind of networks. Here the presence of access points is assumed as ubiquitous, and not shown in the diagram. The inter-BAN communication architecture can be of two types: infrastructure based architecture, and ad hoc based architecture. The outermost communication is the tier-3 communication, where BANs communicate to Cloud via Internet and MSPs, and emergency services such as ambulatory service providers also communicate to Cloud via the Internet. Cloud also initiates communication to BAN, MSPs, and emergency services in case there is medical emergency of MU.

5.4.2 Physical and MAC Layers of BAN

IEEE 802.15.6 standard defines physical layer (PHY) and medium access control layer (MAC) for BAN. It is the task of BAN application developer to develop

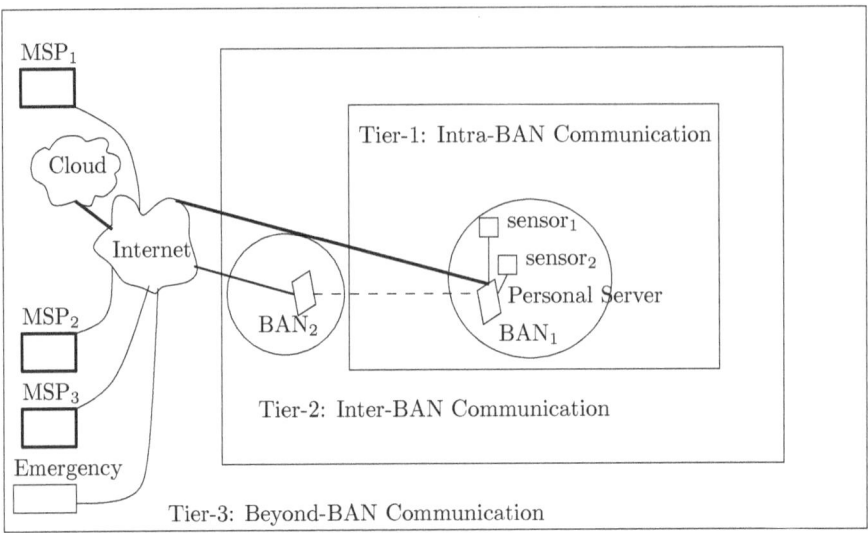

Figure 5.4: Three Tier Communication Architecture

other layers, namely, network layer, transport layer and application layer. PHY and MAC layers are defined so that they provide low cost, low complexity, ultra-low power consumption, high reliability, and short range communication around the body.

Major task of PHY layer is to convert physical layer service data unit (PSDU) into physical layer protocol data unit (PPDU). PHY layer also does the following tasks:

I. activation/deactivation of radio transceiver,

II. data transmit/receive, and

III. clear channel assessment.

IEEE 802.15.6 has mentioned three different types of physical layers: human body communication (PHY), narrow band (NB), and ultra-wide band (UWB). NB PHY is used for communication in the current channel. HBC PHY utilizes the human body, whose tissues are lossy medium, as a transmission channel for transmitting data between nodes [175]. UWB PHY can be used for communication between on-body devices, as well as between on-body and off-body devices.

IEEE 802.15.6 MAC layer controls access of transmission channel. Here, the sensor nodes are organised into one-hop star or two-hop star network. A single coordinator node or hub controls the channel access of the BAN. MAC divides the time axis of channel into super frames or beacon periods of equal length size.

Every super frame has a number of allocation slots for data transmission. The channel is accessible in one of the following three modes:

I. Beacon mode with super frame boundary,

II. Non-beacon mode with super frame boundary, and

III. Non-beacon mode without super frame boundary.

5.5 Cryptographic Building Blocks

In this section, we briefly introduce the important cryptographic building blocks that have been used to address the privacy and security issues in BAN. These include hash function, homomorphic encryption, bilinear pairing, and attribute based encryption.

5.5.1 Cryptographic Hash Function

A hash function maps a input of variable length into a output of fixed length. Hash functions that are used in the security related applications are referred to as cryptographic hash functions [301]. Examples of cryptographic hash functions are MD-5, SHA-1, SHA-2, SHA-3, etc. A function needs to have three properties to qualify as cryptographic hash function. These three properties of hash functions are mentioned below:

I. *Preimage Resistance*: Given a message digest y, it is computationally infeasible to find a message x that hashes to y.

II. *Second Preimage Resistance*: Given a message x, it is computationally infeasible to find a different message $x\prime$, such that both the messages x and $x\prime$ hash to the same message digest.

III. *Collision Resistance*: It is computationally infeasible to find two different messages that hash to the same message digest.

5.5.2 Homomorphic Encryption

Presence of Cloud to process and analysis the MU and MSP demands a different kind of strategy. MU and MSP both do not want Cloud to get access to their actual data, but they want Cloud to be doing some operations on these data. This type of scenario requires that semi-trusted or untrusted servers should operate on encrypted data so that they do not get to know what the actual data is. This is made possible with homomorphic encryption. Homomorphic encryption is a kind of encryption that permits operations on ciphertexts [386]. Homomorphic encryption allows computations on encrypted data, without the need to fully decrypt the data on the Cloud. That is, public Cloud would work on ciphertexts without decrypting it. So, confidentiality of the data is not compromised. Result of applying the operation on encrypted result, when decrypted, matches the result of the operations as if it had been applied on the plaintext. That is, for a particular homomorphic encryption $HE(\cdot)$, and two ciphertexts $HE(m_1)$, $HE(m_2)$, the following holds true: $HE(m_1 + m_2) = HE(m_1) \times HE(m_2)$.

5.5.3   Bilinear Pairing

The basic idea behind pairing-based cryptography, [124], [88], is the pairing between elements of two cryptographic groups and mapping this pairing to a third group, $G_{k} \times G_{k} \rightarrow G_{k}$. For simplicity, only symmetric bilinear pairing is considered, where $G_{k} = G_{k} = G$. Now, we formally introduce bilinear pairing for this simplified version. Suppose, G is a additive cyclic group of order q, and G_T is a multiplicative cyclic group of order q. Here, q is a prime number and let g be a random generator of G_T. A Bilinear map is a non-degenerate and efficiently computable map $e : G \times G \rightarrow G_T$ satisfying the following property:

I. *Bilinear*: $e\left(g^a, h^b\right) = e(g, h)^{ab}$, $\forall g, h \subset G$ and $\forall a, b \in \mathbb{Z}_q$.

Then we say that G and G_T are equipped with a pairing. Bilinear pairing can transform a discrete logarithm from elliptic curve to finite field.

5.5.4   Attribute Based Encryption

In Cloud-enabled BAN, it is natural that only certain users can access vital PHI data. The traditional public key cryptography severely limits the users who can access the content. In attribute-based encryption (ABE), user can encrypt data for a set of receivers who satisfy certain conditions. Here, a ciphertext and a private key are associated with a set of attributes. The key is allowed to decrypt the ciphertext if and only if these sets overlap beyond a certain threshold [294]. There are variations in this basic ABE scheme that support finer-grained access control. In one such scheme, a set of attributes is attached with the ciphertext , whereas an access structure is associated with a private key. This association is specified by a Boolean function. Decryption is possible only when the set satisfies this Boolean function. Since the key specifies the access structure, this scheme is called key-policy ABE (KP-ABE). Its complementary scheme is ciphertext-policy ABE (CP-ABE). In CP-ABE, the ciphertext specifies an access structure whereas a key is associated with a set of attributes.

5.6   Privacy and Security

Any system connected to Internet and Cloud must take adequate measures for ensuring security and privacy. The Cloud-enabled BAN is also no exception. Rather, privacy and security is of paramount importance in Cloud-enabled BAN. This is because any compromise of the PHI parameters, MSP parameters, and location of MU and MSP will lead to failure of the system.

5.6.1   Security Notions in Cloud-enabled BAN

Cloud-Enabled BAN must satisfy some of the following security requirements:

I. *Data Privacy*: The PHI data of MU is very sensitive. MU does not want her personal PHI data that depicts her physical status to be known outside. Neither attacker nor Cloud should be able to get access to this raw data. However, those data are encrypted and then processed by Cloud and MSP because any compromise or alteration of MU PHI data can lead to fatal consequences such as wrong medical treatment or false medical emergency alarm generation, etc. Similarly, MSP also does not want to disclose its infrastructural parameter details to the outside world. However, Cloud can access MSP data in encrypted form for processing.

II. *Identity Privacy* or *Anonymity*: Neither MU nor MSP want to disclose their real identity to a third party such as attacker or semi-trusted Cloud. If real identity of MU gets disclosed then that can be life threatening to those who are very important persons or someone in society whose physical status on diseases can lead to loss of market share, brand values, etc. Similarly, leakage of identity of MSP can be exploited by its rival MSPs.

III. *Location Privacy*: Location of MU is very important parameter that MU does not want to share with others. Location information of MU can be exploited by others if disclosed by entities like Cloud.

IV. *Authentication*: Since multiple entities like MU, MSP, and Cloud exchange data, their source must be authenticated properly by the system.

V. *Unlinkability*: Adversarial agents must not be able to link two messages generated from the same BAN.

VI. *Access Control*: As different types of users might be there who process the PHI data, there should be differential access rights to process PHI data. For example, the emergency medical technicians should access only the data during a medical emergency whereas the assigned doctors must access past records to find any correlation that can help in proper medical diagnosis.

5.6.2   Attacks and Threats in Cloud-enabled BAN

The attacks and threats are generally classified into passive and active. Passive attack does not modify the transmitted data whereas an active attack modifies the intercepted messages. Following are some common attacks on BANs:

I. *Eavesdropping attack*: Since BAN communication takes place wirelessly, it is not difficult for an adversary to listen to the data communicated through an insecure channel. The goal of this passive attack is to know the content of the communication by collecting a large amount of exchanged messages. It is very important to stop this, as PHI data is very personal. To prevent eavesdropping, PHI data has to be encrypted and the keys used must be updated regularly.

II. *Impersonation attack*: This attack enables an attacker to pose itself as an authorized user of the network. The attacker gains private identity information and encases it to impersonate a legitimate user in the network. As a result, the attacker obtains the same rights as of the legitimate user. Authentication mechanism needs to be adopted by different entities in BAN to protect from this kind of attack.

III. *Replay attack*: An attacker captures a message sent by user and resends it at some later time. If appropriate countermeasures are not present, this resent message is accepted by the receiver as a valid message. The attacker can achieve some malicious intent. Since the receiver processes the message, receiver can be misled by the message content. Also sensor energy gets reduced if plenty of non-legitimate replayed messages are processed by the legitimate user. To prevent this attack, the BAN communications should be secured in such a way that it stops old messages from being sent again at a later time. For achieving this, authentication and message freshness are to be ensured in BAN. This is often done by using timestamps or nonces.

IV. *Man in the Middle attack*: In this attack, the attacker intercepts messages being sent by sender, modifies the messages and then relays the messages to the receiver. Sender and receiver believes that they are communicating with each other directly. This powerful attack allows the attacker to eavesdrop and manipulate the transmitted message in real time.

V. *Denial of Service Attack*: Denial of Service (DoS) is an attack that targets to disrupt the availability of certain resources or services. It is made possible by flooding the target with a huge number of fake messages until the target entity is no longer able to process them all. Besides degradation of performance services, energy of BAN nodes is drained heavily by this type of attack. Defending against this attack is done mainly by trying to identify legitimate messages ignoring the suspicious messages.

5.6.3   Existing Security and Privacy Solutions in Cloud-enabled BAN

This section presents different solutions proposed in the literature to ensure privacy and security in Cloud-enabled BAN. [223] proposed a privacy-preserving emergency notification scheme to the nearest doctor through mobile healthcare social network. When medical emergency occurs, the personal digital assistant (PDA) of MU accumulates the emergency data such as PHI record, MU health record, and MU location. It then broadcasts this emergency call along with emergency data through mobile healthcare social network. Thus the message is communicated to the nearest doctor. It has adopted attribute-based encryption (ABE) scheme to make sure about proper access-control of emergency data. It has also discussed the revocation of the access control for maintaining safety from insider attack.

The work proposed in [345] considered separation of access rights for the health records between emergency medical technician (EMT) and primary physician. Primary physician is fully trusted, whereas EMT is considered as honest-but-curious. Remote server is considered as semi-trusted. The EMT, remote server, and credential authority may collude to compromise patient's privacy. This work allows EMT to get access to only those data that are required for addressing the medical emergency, whereas primary physician is allowed to get unrestricted access to all the past health records of MU. The different security privacy concerns that have been addressed in this work are authentication, anonymity, location privacy and unlinkability of multiple medical data of same MU but sent at different time. It has used cryptographic primitives such as bilinear pairing, commitment scheme, proof of knowledge, anonymous credentials, and pseudo-random number generator (PRNG) for achieving the privacy and access control.

[91] has developed an event-aided packet forwarding protocol that enables patients to communicate with each other in mobile healthcare social network (MHSN), built upon BAN and mobile communication platform, whenever illness related events or activities occur. They have used predicate encryption to provide patient identity privacy, patient illness privacy and message confidentiality.

Another work considered privacy-preserving and secret sharing of data and collaboration scheme in MHSN of smart cities by [172]. They have used identity-based broadcast encryption and ABE scheme for secure and privacy-preserving social and health data sharing, respectively. Further, to provide a secure data collaboration from independent cloud servers, they have used proxy re-encryption (PRE) scheme. As a result, the authorized health data analysers can access re-encrypted data.

Tackling of infectious disease in smart city has been explored in [396], which demands fusion of health cloud data and social network data analysis. This is a promising area, which has not been explored much up until this date. [236] has considered a different scenario in MHSN, where a patient can share its data to other members who also have the same symptoms. They have proposed a secure same symptom based handshaking scheme using bilinear pairing.

After sensing of PHI data by sensors in the BAN, the data read by the sensors may be noisy due to reasons such as lossy human body medium and uncertainty of MU motion. To remove these noise, data has to be filtered. [216] has proposed the unscented Kalman filter based emergency fusion model to increase the data accuracy when emergency occurs.

So far all the works we have mentioned focussed on content privacy. Content privacy is all about protecting the MU information privacy. Whereas [73] has considered both content privacy and contextual privacy. Context privacy is about the communication context and it has four parts: sender/receiver anonymity, un-linkability, un-observability and pseudonymity. It has proposed contextual privacy based on the concept of onion routing, fake message injection and multi-cast, whereas it has used identity based cryptography to provide content privacy.

Another work by [96] focussed on attending of MU in emergency by nearest available medical personnel. During medical emergency, when no one is around to help, their system enables the nearest medical personnel to be alerted so that the MU is attended at the earliest. The Cloud, aware of the emergency event, grants necessary access to medical data of the patient to the medical personnel heading towards the MU in emergency. This system has provision for cancel phase in case someone nearby arrives to help the MU.

A lightweight encryption algorithm based on SHA-3 has been proposed by [385] for secure communication between BAN and server. Apart from keeping the data in servers safe, it employs Sharemind, a framework easy for use by non-crypto users. Sharemind splits the patient data into three numbers such that its sum equals the original data. As an improvement to this work, [363] supports any number of participants and is secure till half of the participants are not compromised. This proposal uses another lightweight multi-party computation protocol based on FairplayMP framework. This scheme has an advantage that a sensor needs to keep just one key to communicate with n number of different servers.

Since the BAN nodes operate in lossy medium, it is very natural that some data readings will be faulty. [190] has attempted to identify the data faults and reconstruct them in pervasive healthcare. It considers that there exists correlation in multiple attributes among different sensor nodes in the MU. Any activity is reflected in the change of readings of at least two sensors. Here every node maintains a trust rating for other nodes using Cosine similarity.

Attribute based encryption has been used by many works for providing privacy and security in BAN. ABE has the advantage that it is capable of providing differential access rights for different stake holders in BAN-centric system. There are two flavours of ABE: Key-Policy ABE (KP-ABE) and Ciphertext-Policy ABE (CP-ABE). [352] studied the suitability of these two approaches for application in BAN and found that KP-ABE is more preferable.

The problem of data privacy in Cloud-assisted healthcare systems has been reviewed by [320]. An outstanding work on Cloud-assisted privacy preserving mobile health monitoring is by [225]. It has used outsourcing decryption technique and multi-dimensional range query to shift decryption complexity from

MU to Cloud. They have further used key private proxy re-encryption to reduce computation complexity of service providers. A systematic literature review on distributed denial of service attack in Cloud-assisted BAN has been carried out by [212]. [360] has proposed a model for Cloud-assisted mobile-access of health data where both private Cloud and public Cloud have been used. Data processing and analysis tasks are done at private Cloud and processed results are stored in public Cloud.

Privacy-preserving priority based data aggregation for Cloud-assisted BAN has been discussed in [395]. The health data is categorized into different types and accordingly assigned different priorities. Then separate forwarding strategies are chosen as per the data priority. They have used bilinear pairing and Paillier cryptosystem for achieving privacy-preserving data aggregation.

5.7 Authentication in BAN

Authentication is one of the desirable properties in secure systems. As communication is mostly wireless in Cloud-enabled BAN, the source authentication is very important. [217] has proposed a secure authentication scheme using Chebyshev chaotic maps for Cloud-assisted BAN. In this work, the MSP to treat MU is pre-decided by MU herself.

[378] proposed a revocable certificateless encryption scheme and revocable certificateless signature scheme to provide certificateless remote anonymous authentication protocol. The salient feature is that the revocation scheme is scalable. [151] proposed a distributed attribute-based authentication system, where MU/MSP use their verifiable attributes for authenticating each other. It also provides privacy protection and verifiability of MU/MSP attributes.

[229] has proposed an anonymous authentication protocol that used an anonymous account index instead of actual identity of MU to access BAN service, thereby preventing the potential privacy leakage. However, [377] and [161] proved that this scheme is insecure against public key replacement attack, and impersonation attack. [377] has proposed a lightweight certificateless scalable and remote anonymous authentication protocol by using the certificateless encryption and two-party authenticated key agreement protocol. An excellent work on anonymous authentication in BAN is by [161]. They have proved that their anonymous authentication scheme is mutual authentication-secure based on the hardness of Diffie Hellman problem.

5.8   Key Management in BAN

Since the data in BANs are transmitted through wireless medium, the communication has to be secured, to prevent eavesdropping and distortion PHI data. For this, a cryptographic scheme needs to be employed, and that requires secret keys. That is why it is necessary to have a secure key agreement and distribution of them amongst the sensor nodes in the BAN. As sensors in BANs have very limited resources, it is important for the key agreement scheme to be lightweight. We can classify the existing key agreement schemes into four types [196], [33]:

   I. Traditional key agreement schemes

   II. Physiological value-based key agreement schemes

   III. Hybrid key agreement schemes

   IV. Secret key generation schemes

The traditional key agreement schemes are generally based on public-key cryptography or some kind of key pre-distribution techniques. This requires more key storage space, but reduces the processing effort. The second type of key agreement schemes are physiological value-based key agreement schemes. These use humans' vital physiological characteristics such as ECG signal and then transform these values into a secret key by adopting some appropriate artificial intelligence technique. Since the sender and the receiver are in the same human body, they generate the same key. This removes the need for key pre-distribution and explicit key distribution. The third kind of scheme is the hybrid scheme, which uses both of the previous kinds of key agreement approaches. For the key generation and agreement, they often use key pre-distribution and physiological values from the human body. For an introduction to key pre-distribution, one can refer to [302]. Lastly, the secret key generation approach is mainly based on the characteristics of communication channel and physiological data from the user. These signals are generally used for independent key generation of a key, while physiological signals are used for key exchange. [119] has proposed a key management scheme for group communication in BAN by using physical unclonable function (PUF). Group communication in BAN is an area which has not been explored much.

5.9   Conclusion

In this chapter, we have discussed the Cloud-enabled Body Area Network technology. This is a promising area of research with so many critical issues involved. We have attempted to introduce the sensor nodes, bio-sensors, and the body area network gradually. The architecture of BAN and Cloud-enabled BAN has been presented. Although there are many applications of BAN in many path-breaking domains, we have focussed our discussions on mainly application of BAN in healthcare, particularly in continuous healthcare monitoring. Since BAN enabled medical user's physiological data and location information is extremely private information, any BAN-based system will not succeed if proper security and privacy measures are not taken. This privacy and security issue is more relevant in the presence of Cloud. Cloud-enabled BAN demands that Cloud will be processing the medical user health data and location information in encrypted form so that it can not extract any contextual meaning of BAN data. We have presented the security and privacy issues in Cloud-enabled BAN and discussed some of the research done in this direction. Also, we have very briefly discussed the key agreement in BAN. With the advent of Internet of Things, this horizon of Cloud-enabled Body Area Network is going to enormously expand in the near future.

Chapter 6

Trust and Access Controls in IoT to Avoid Malicious Activity

Yenumula B. Reddy

Department of Computer Science, Grambling State University, Grambling, LA 71245, USA

Shahram Latifi

Dept. of Electrical and Computer Eng., University of Nevada, Las Vegas 89154, USA

Connect and control access to the Internet of Things (IoT) is crucial since it converges and evolves multiple technologies. Traditional embedded systems, wireless sensor networks, control systems, software defined radio technologies, and smart systems contribute to the Internet of Things. The deep learning, data analytics, and consumer applications have an essential role in IoT. The challenges are to store, process, and develop meaningful form so that it is useful to the business, government, and customers. The chapter discusses computing data in a cloud, current challenges to store, retrieve, process, as well as security requirements, and possible solutions. We frame the trust-based access control model that incorporates a digital signature to minimize the malicious activity for storage, retrieval, updates, and processing of sensitive data. We further provide the trust framework for the user and access control algorithms for data processing in the cloud environment.

6.1   Introduction

Storing, processing, and retrieving information on a cloud is not straightforward. Today, small devices are involved in most of the information processing. The performance and security in transferring the data in small devices do not meet such requirements (storing, processing, and retrieving) due to their low computational power capabilities, particularly in small and moderate power-operated devices. Adding the cloud capabilities to these devices eliminate these constraints. Further, the data (sensitive or non-sensitive) in the cloud may be in an encrypted format. The encryption technology is an expensive task in small devices (local operations and processing) due to the computational resource requirements and processing. If a device connects to the cloud, most of the transactions will be done in the cloud. At the cloud level, in addition to encryption and decryption, the user requires to search, update, and process the stored data. Further, most of the operations (read, write, edit, and process) of sensitive data requires encryption and decryption process due to the nature of the data. These operations are not practical in IoT devices in the currently available technology.

The Internet of Things (IoT) consists of physical objects embedded with sensors, software, cloud (computing facilities distributed over various sites), and network connectivity of these objects to store, exchange, process, and decisions on complex data. In smart cities, the data generates from various activities, events, and other sources. Most of the information is general-purpose, and some data will be useful for decision making. Creating or generating such data from various sources (mostly from sensors) and filtering the data is part of the big data analysis that will be used for decision making on a real-time basis. Some of the sources the data will be generated through sensors may be traffic signals, smart devices, embedded systems, and traffic. These data create in large volume, continuously, and high velocity. Technically, such data is called big data in current technology, and analysis of big data is very important for decision making on a real-time basis.

The word big data was coined recently and became famous due to its volume (beyond the size of the standard database), velocity, and variety (numeric, text, images, videos, and combination) [287], [26]. Most important point is its capture, storage, analysis, and production of useful results for decision making. The explanation goes beyond the previous statement. It is an unstructured, vast volume, continuously growing on a real-time basis, and a challenging problem to make it useful or filter and obtain the most valuable data for decision making. The technology changes influence the classification changes of such data over a period. Therefore, its definition changes time to time and organization to organization (challenging to have a perfect description). Every organization has vested interest in the classification of such unstructured data. As technology changed from computers to hand devices, the processing became a big problem. Due to this reason, cloud requirement exists, and the cloud is the future. Further, cloud technology helps to take most of the storage of a large volume of data, complex computations, and generation of customer output.

Cloud requirement for big data is due to its size, a different type of data (heterogeneous) from multiple sources, the velocity of its flow (incoming and outgoing), potential value if adequately classified and processed, and confidentiality [26], [180]. Cloud computing is a paradigm with unlimited on-demand services that is software as a service (SAAS). It can virtualize hardware and software resources, with high processing power, storage, and pay-per-usage. Moreover, it transfers cost calculation responsibilities to the provider and minimizes the exceptional setup of computing facilities at small enterprises. It has negotiable natural resources and gets computing power as required. The cloud services provide infrastructure as a service, software as a service and platform as a service.

Since the Government, business, and education are moving into the cloud, the requirement for experts in cloud computing continuously increases. There will be a big vacuum in this area soon. In recent years, the cloud infrastructure as a service (IaaS) market has shaken out contenders. The superpowers in technology include Amazon Web Services, Microsoft Azure, and Google Cloud. These organizations sit above providers, vying to carve out a niche in technology with a prohibitively high cost of entry. Therefore, the demand for cloud computing solutions is sky-high. The infrastructure as a service (IaaS) market is currently growing at more than 23%, compared to the overall cloud market growth of 13.8%. Therefore, businesses must foresee the future and utilize the increased flexibility and scalability of the cloud infrastructure and what it offers over traditional on-premises solutions

Cloud computing delivers storing and processing of sizeable unstructured volume of continuously generated data, resource availability, and fault tolerance through its various hardware and software facilities. Many companies, including Nokia, RedBus, Google, IBM, Amazon, and Microsoft, provide consumers to consume service on-demand. Big business decided to migrate to the Hadoop distributed file system (HDFS) that integrates data into the same domain, using supplicated algorithms to get proper results for its customers. The advantage of using Hadoop is cheaper storage compared to traditional databases. Currently, HDFS helps Nokia, RedBus, Google, and other companies to fulfill their needs. The facility helps these companies to concentrate on their businesses rather than on technical details and requirements.

Big Data technology solves many problems irrespective of volume, velocity, and source of generation. It is a continually changing technology, and many industries, customers, and government agencies are involved in usage and managing. Further, the data is in a cloud environment. Due to this reason, we need to create security policies, access rights, secure storage, and retrieval. Even though data continuously grows, controlling is required (means of access to valuable data that requires security). Therefore, we need to enforce data governance policies like organizational practices, operational practices, and relational practices [157], [399].

Disaster recovery (in the case of dangerous accidents, including floods, earthquakes, fire, and accidental loss of data) for valuable data is a requirement. The big corporations define a set of procedures for a disaster recovery plan to restore the data. In addition to security policies, disaster recovery is strongly recommended (fault-tolerant depends on disaster recovery). The other problems include the secure transfer of data to the cloud, incorporating high-performance computing, and data management. Big data in the cloud has many research and practical challenges. Storing the data using an encryption technique takes extra time with standardization of procedures to minimize the impact of heterogeneous data. Data governance, recovery plans, quality of services for secure transfer of data, and petaflop computing are some of the problems for implementation.

IoT generates a significant amount of complex data from multiple sources and requires processing real-time basis with defined policies and privacy requirement. There is a need to represent high-level aggregating requests and often involves inference techniques. This data needs various controls, human intervention, and feedback. Also, there is a need to identify the devices, customers, and access limits (control level). The restrictions include data access and retrieval (inside the system model and hierarchical control). Besides, the customer connecting device must be trusted to eliminate malicious activity. That means, each entry and exit of device or resource needs to be logged and there needs to be analysis of log for security decisions and actions.

The successful deployment of complex data activity on a cloud requires building a business case with an appropriate strategic plan to use the cloud. The project must develop productivity, extracting more significant value, continuous improvement, customer acquisition, satisfaction, loyalty, and security, as well as assess suitable cloud environment (private or public) and develop a technical approach. Next, address governance, privacy, and security increase risk and accountability requirements. Finally, deploy the operational environment. The provider meets many challenges depending on the cloud data environment. Security, cost factor, customer satisfaction, and service reliability are central issues.

6.2   Threats, Vulnerabilities, and Access control Requirement in Internet of Things

6.2.1   Threats

Currently more objects are continuously connecting and disconnecting to the Internet than at any time before. The users have many mobile adoptions decentralized with various operating environments that encourage the attack. Users (customers) individually or in a group linked to a non-linear number of direct and indirect entities (heterogeneous devices) increase the complexity and make it difficult to identify the source of an attack.

There are many characteristics favorable to attackers. The first concern is many users geometrically increasing with the diversified operating environment. The user demands to increase the requirement of storage, battery life, and real-time response with various resource options (encryption, video transferring, text messages, and audio requirement) in addition to software updates for devices and Apps. Music and games will take total resources including spectrum requirement, processing, and transferring of information. Further, the connection of heterogeneous devices and interoperability between entities within and outside the environment are dynamic due to their characteristic properties. Another complex relationship to the network is sensors and continuous transfer of information.

Attackers find the characteristics of IoTs controlled by the technology world and use specific tools to achieve their goals. Due to this reason, we require the trust of a user and access limits that minimize the threat.

6.2.2 Vulnerabilities

The devices that interconnect real-world sensors including wireless sensor networks (healthcare monitoring systems, weather monitoring systems, data logging, industrial monitoring, and environmental monitoring systems), internet connecting vehicles, smartphones, fitness monitoring devices, and tablets are a few examples of IoTs. The main IoT enabling technologies are radio frequency identification, wireless sensor networks, and cloud computing. IoTs are structured in three layers: a physical layer, network layer, and application layer. Hackers can penetrate through any of these layers. Vulnerability assessment of these devices and software supporting these devices are essential to trust the functionality of these devices. Trusting the feature of these devices requires four steps.

1. Identify the leading causes of those successful attacks, vulnerabilities, the commonalities, and the knowledge to improve the security of embedded systems. Use a commercially available field-programmable gate array (FPGA) board to develop the firmware-level vulnerability assessment and the application-level vulnerability assessment algorithm.

2. Develop a Vulnerability Assessment (VA) framework, and conduct a penetrating test on a small scale embedded system (network system) and collect scan logs.

3. Analyze the scan logs based on machine learning algorithms.

4. Develop an automated VA prototype and test the simulations.

6.2.3   The importance of Access controls and Trust of users

Cybercriminals focus on IoT devices with various techniques including social engineering, malware, and different hacking methods. Once a device touches the network, the issues including patching, monitoring, and quarantining require to establish the visibility. These devices are notoriously vulnerable and impossible to secure in today's situations. Due to increase in number every day, the organizations are not able to keep an inventory. The first technology in network access is access control. Once the user enters into the network, the user access to resources will be established. It depends on the type of user, device, time, place, and resource limitation. Clear policies must be developed and implemented. There are many reasons to incorporate the access controls as part of user authentication.

1. Cyber-criminals are actively increasing their focus on IoT devices, with the latest variant of hiding seek malware expanding its focus to include, for the first time, home automation devices.

2. The IoT devices are notoriously vulnerable to attack

3. Most organizations have no way to track these devices since thousands of new tools are added every day

4. These devices do not have resources to adopt at the speed the devices are increasing

5. So we need a particular model to assess the vulnerability of these devices

6. The assessment will then lead to control the unauthorized access to these devices

In the access control mechanism, we have to follow a few important policies.

1. User managed access and authorization layer incorporated in communication protocol (developed by IBM, Object Management group, user management access (UMA), and internet based services (need to confirm quality and reliability)

2. Trust-based mechanism

3. The connection process requires authorization

Identifying user and authenticating is one of the requirements before assigning controls. Figure 6.1 has three levels that includes IoT device, access control server, and internet connection. Once the device activated, it connects to an access control mechanism that requires to open the device access to resources. Since the number of devices connected and disconnected change continuously to the internet, normal access control mechanisms will not work. It requires a special mechanism to limit the resources depending on type of user. Figure 6.1 shows the example of the mechanism.

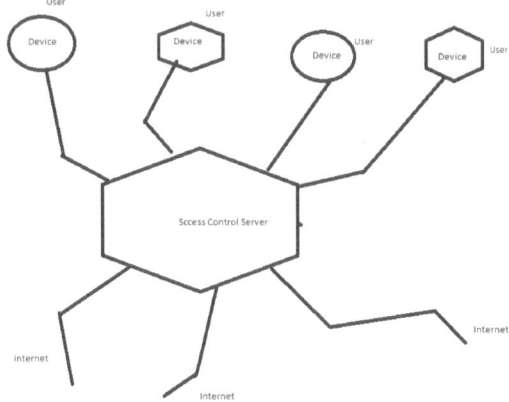

Figure 6.1: Basic Access control Access

6.3 Literature Review

Storage, processing, and retrieval of big data in the cloud are significant problems in current research. Pedro et al. [287] studied the overview of present and future issues. The document discusses scalability and fault tolerance of various vendors including Google, IBM, Nokia, and RedBus. The authors further considered the security, privacy, integrity, disaster recovery, and fault tolerant issues. The authors [2] discussed the review of current service models, import concepts of cloud computing, and processing of big data. Elmustafa and Rashid [26] presented the survey issues of big data security in cloud computing.

Linda et al. [180] showed the environmental examples of big data use in government that includes Environmental Protection Agency, Department of the Interior, Department of Energy, and Postal Services. The study consists of the government open access initiatives, federal data center consolidation initiative, and the enforcement of compliance online. James [195] presented a roadmap to the success of big data analytics and applications. The report discusses the definition and description of unstructured data, relevant use cases in the cloud, potential benefits, and challenges associated with deploying in the cloud.

The impact of cloud computing on Healthcare is studied in [290]. The study includes on-demand access to computing and large storage, supporting big data sets for electronic health records, and the ability to analyze and track the health records. Alan presented the environmental sustainability of big data, barriers, and opportunities. It also includes new opportunities for partnership based collaboration, sustainability to organizations to big data efforts, and emerging business models.

Yan et al. [391] discussed the access control in cloud computing. The paper contains the temporal access control in cloud computing using encryption techniques. Yuhong et al. [230] discussed data confidentiality in cloud computing. The article uses the trust-based evaluation encryption model. In this model, the trust factor decides the access control of user status. Young et al. [265] discussed the security issues in cloud computing. The paper describes the access control requirements, authentication and ID management in the cloud.

Ali and Erwin [140] reviewed security and privacy issues on big data and cloud aspects. They concluded that cloud data privacy and safety is based on the cloud provider. They also discussed big data security challenges and cloud security challenges. Their paper examines the security policy management and big data infrastructure and programming models. They did not suggest any particular model but discussed all possible solutions for the security of big data in the cloud.

Marcos et al. [51] presented approaches and environments to carry out big data computing in the cloud. The paper discusses the visualization and user interaction, model building, and data management. Venkata et al. [177] examined issues in a cloud environment for big data. The primary focus is security problems and possible solutions. Further, they discussed MapReduce and Apache environments in the cloud and the need for security.

Saranya and Kumar [322] addressed the security issues associated with big data in a cloud environment. They suggested a few approaches for the complicated business environment. The paper discusses unstructured big data characteristics, analytics, Hadoop architecture, and real-time big data analytics. The authors did not present any particular model in the article. They explained a few concepts related to security in a cloud environment.

Avodele et al. [55] presented issues and challenges for deployments of big data in the cloud. They suggested solutions that are relevant to organizations to deploy the data in the cloud. The authors indicated the importance of authentication controls and access controls.

Security in IoT was discussed in [249], [331]. The authors conclude that IoT networks are hugely needed to ensure confidentiality, authentication, access control, and integrity, among others. The reason for immediate attention to security is a dramatic increase in the number of connected devices. These devices create technical problems such as attacks with a broader scope of influence and attacks that last longer. Therefore, immediate attention and procedures are required to detect the hacker to avoid the damage to sensitive information.

Zhang and Wu [399] discussed the generation of data as trustworthy or not. Further, the paper focused on access control models on trust computing and useful guidelines. Dina et al. [174] investigated the requirements in IoT to a candidate's vision. The authors thought the access control should be implemented during the requirements stage. Ali et al. [35] discussed the device authentication and access control in IoT. The put the efforts to demonstrate the security requirement to eliminate the possible cyber-attacks. Situational access control in IoT was discussed by Roei et al. [85]. They identified that the situational tracking requires cross-framework interaction and permission. In this process the system tracks to sense the situation, infer the situation environment and then activate the process.

The remaining paper discusses the problem formulation that leads to the authentication model in the cloud section 6.4, simulations in section 6.5, access control algorithms in section 6.6, and conclusions and future work in section 6.7.

6.4   Problem formation

The security model involves the cloud customer data security at storage, retrieval, transfer, processing, and updates (insert, modify, delete). The security needs to be set at the log entry at user and cloud level. It also requires the automatic validation of stored data status and verifies the trust level. The framework of the proposed model includes the data encryption, correctness, and processing. These three modes depend upon the access rights of the user as discussed at the beginning of the current section. For storage and retrieval of data the basic encryption techniques, Advanced Encryption Standards (AES), Rivest-Shamir-Adleman (RSA), and steganography model are sufficient. If the data requires storage and processing the recommendations in [106] may be useful. The paper discussed the various techniques to search cipher text and query isolation (avoid the untrusted server). Controlled searching, dealing with variable word lengths, searching encrypted index, and support for hidden search are part of the research. In this paper, the proposed access control model with encrypted processing data is useful to avoid the untrusted provider and malicious users in the cloud.

The access control model for IoT of cloud storage incorporates the authentication of customer and its current access level. The token identification (TID) is attached as soon as the user logs in into the system. To maintain the security of data and its trust level, we have to define many control parameters to the user access in the cloud. The current TID model in equation 6.1 explains with seven parameters.

$$TID = UID, IID, MDT, TA, PA, LGE, SA \qquad (6.1)$$

where

UID User Identification and access rights
IID Issue Date
MDT Maximum Date (expiration Date)
TA Time of Access
PA Place of Access (current place and node ID)

LGE   Log Entry (*UID, IID, MDT, TA, PA, LGE*)
SA   Security Alarm

The customer is an owner of the data or another customer. In either case, the customer is a client with different access rights. Once the customer logs in to the cloud network, the authentication access token connects to the user account. The token verifies the customer/user access limits and allows or denies appropriate file access. Further, the system does the entries in the customer/user and cloud log table for each attempt of a user to a particular file with all details. The various validating and verification checking the modifications help to find unauthorized access. The trustworthiness of provider or customer/user can be calculated using the log values.

The trustworthiness of a customer/user can be calculated using trust function in equation 6.2. For each entry of the user, the weight 'W' is assigned. The entry $W_{i,j}$ means i^{th} user and j^{th} entry. Let N_g be the number of times the user has right behavior, and N_b is the number of times of bad behavior of the ith user. Multiply the user entry value with weight 'W' with right or wrong actions and calculate the trustworthiness of a customer/user. The trustworthiness T_i of ith user/customer is calculated as follows.

$$T_i = \frac{\sum_{i,j} W_{i,j} * Ng_j + x}{(\sum_{i,j} W_{i,j} * Ng_j + x) + (\sum_{i,k} W_{i,k} Nb_k + y)} \tag{6.2}$$

If T_i the trust value is above the threshold, the customer/user is considered good; otherwise a false alarm alerts the owner. The user may be a customer, provider, or owner. The weight varies between 0 to 1 and the number of times there is access to the data (or data files) will be 0 to 10. If weight $=0$ then $x=1$ else $x=0$. Similarly, if the number of times $=0$ then y value is 1 otherwise $y=0$.

6.4.1   Improved Trust calculation

The improved trustworthiness of the user can be calculated as shown in equation 6.3. The equation 6.3 estimates the trustworthiness T_i of the ith user.

$$T_i = T_i + \frac{\sum_{i,j} W_{i,j} * Ng_j + x}{(\sum_{i,j} W_{i,j} * Ng_j + x) + (\sum_{i,k} W_{i,k} Nb_k + y)} \tag{6.3}$$

Initially, T_i is set to 0. We provide the initial values to calculate the first T_i. If T_i, the trust value, is below the set threshold, the user is considered bad and signals the false alarm. The user may be a customer, provider, or owner. The weight varies between 0 to 1 and the number of times the data (or data files) has been accessed will be 0 to 10. If weight $=0$ then $x=1$ else $x=0$. Similarly, if number of times the cloud has been accessed $= 0$ then y value is 1 otherwise $y=0$.

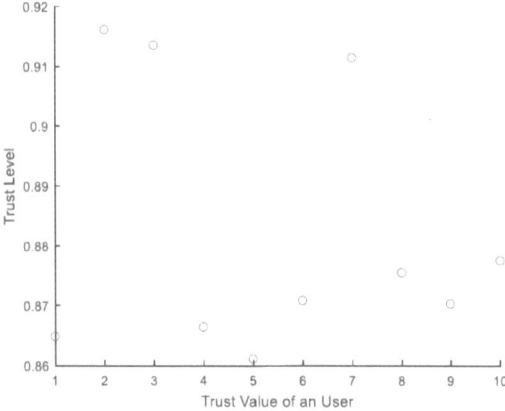

Figure 6.2: Experiment 2: Average of 10 random accesses of a user to data by each user

6.5   Simulations

The simulations on equation 6.2 were performed and provided in Figures 6.2, 6.3, and 6.4. The threshold value for legal (trusted) user was fixed at 0.85 and above (>= 0.85) in the current situation. We assumed that every user logged in to the internet is not a completely trusted user (may be hacker or malicious user) at any assumed average trusted level. For example, if the threshold value is greater than 0.85 on an average ten (10) attempts, then the user is legal (assumption) or trusted. The program was developed in MATLAB language to create a graph for equation 6.2. Figures 6.2, 6.3, and 6.4 present the sample results. The random data generated for each user and plotted the average of 10 access values. Figure 6.3 shows that the user is malicious or untrusted based on the number of attempts to the system or trying to access to specific database or information. The result automatically triggers the alarm to security manager. That means the trust level of user accesses calculated is below 0.85. In our simulations, the user access values depend upon the random values and the corresponding weights selected provided during the execution. The proposed data is a random selected sample for the test calculation. The complex calculation requires a real data (not provided) since it has various parameters for each user logging and processing the data.

The improved trustworthiness of user is calculated as in Figures 6.5, 6.6, and 6.7. The threshold value depends on a selection that the user is legal or malicious. As in the previous case, if the user is genuine, the threshold value is higher than 0.85 for an average access of a user to cloud equals ten (10). The simulations are created using MATLAB and results are shown in Figures 6.5, 6.6,

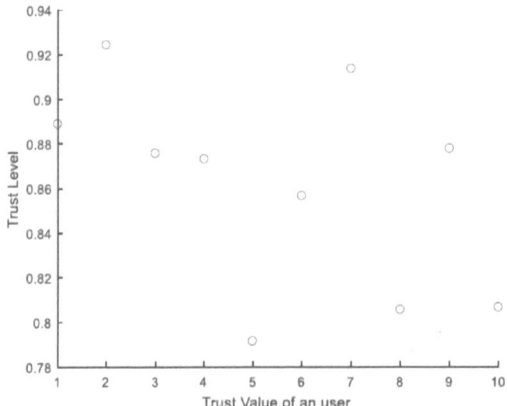

Figure 6.3: Experiment 3: Average of 10 random accesses of a user to data by each user (sample graphs)

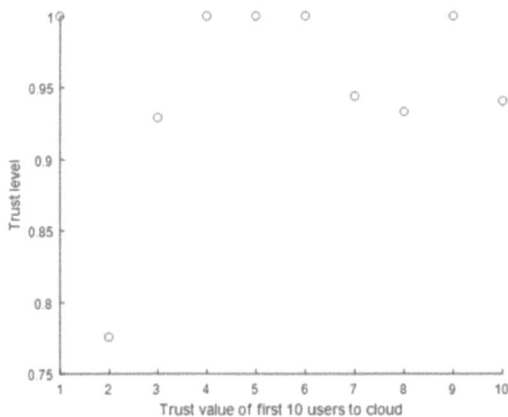

Figure 6.4: Experiment 4: Average of 10 random accesses of a user to data by each user (sample graphs)

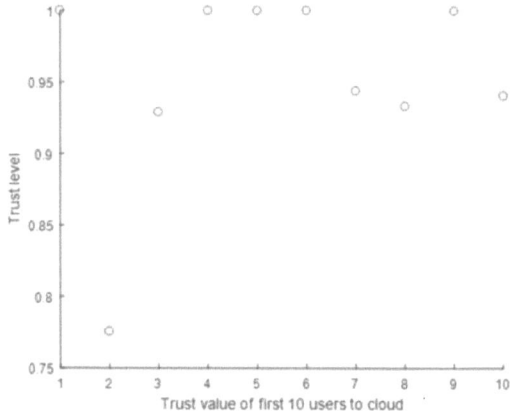

Figure 6.5: Experiment 5: Average of 10 random accesses of a user to data by each user (sample graphs)

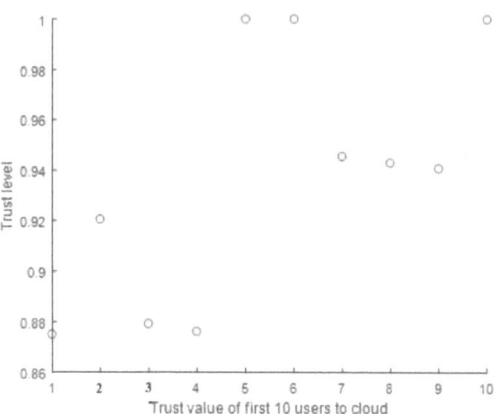

Figure 6.6: Experiment 6: Average of 10 random accesses of a user to data by each user (sample graphs)

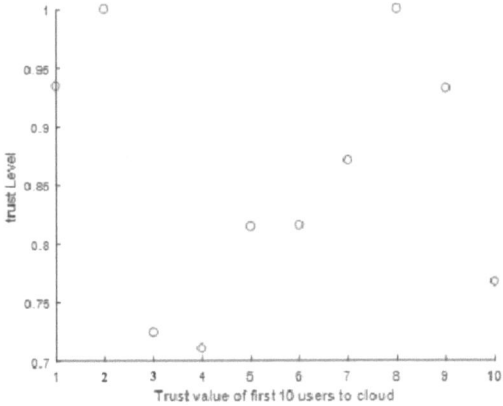

Figure 6.7: Experiment 7: Average of 10 random accesses of a user to data by each user (sample graphs)

and 6.7. The random data generated for each user and plotted the average of 10 access values. The figures indicate zero or more malicious users. Figure 6.7 shows more hacking activities compared to Figure 6.5. Note that the proposed data is a simple test calculation.

6.6 Access Controls on Sensitive Data

Access to sensitive data cannot satisfy pure trustworthiness. Along with trustworthiness, the procedure requires the user access limits, day, time of the day, and log entry for validation. The user identification contains access rights and *UID* issue date, expiration date, time of access, and location of access (depends upon sensitiveness of data). Once the user logs in to the system, the cloud log, and owner log's entries are automatically registered. For hackers, only cloud log entry appears. The various validation and verification checks reveal the hacking. The token identification parameters in equation 6.1 are used in objective function *G*.

$$G = \{N, A, D, U\} \qquad (6.4)$$

The objective function *G* replaces *TID*, *N* replaces *UID* (contains *IID*, *MDT*, *TA*, and *PA*), *D* is a data file (or database), and *U* replaces *LGE*. The security alarm will be activated depending upon the hacker identification or trust failure. Therefore the parameters are explained further as below.

N the set of users $(n_1, n_2, ..., n_m)$

A set of access rights $(a_1, a_2, ..., a_p)$

D set of allowed resources in file or database $(d_1, d_2, ..., d_q)$

U the result of the query and log entries for verification and validation.

Once the authenticated user $n_i(n_i \in N)$ logs into cloud environment, the CCCRN service attaches a_i service token to a resource within its domain with a set of access types. The limitation helps to control the user for resource access. For every service requested by the user, the system generates a set of access permissions to the resources. The services required should not exceed the user access limits. If the resource requirements are outside the user boundaries, then the system alarms the security and denies the request. Hacker is a user that does not have any role in the system. An authorized user will be treated as a hacker if the user tries to access unauthorized information. For example, the healthcare staff member will be considered as an intruder if the user accesses unauthorized data or misuses (for instance, printing and forwarding) the authorized information.

In the proposed CCCRN environment the user with complete authorization access is called a super user (S). The super user 'S' possesses access rights of all users $S \supseteq \underset{i=1,n}{a_i}$ where $\supseteq$ means contains. All accesses of super user on the database must be recorded. The user that does not have authorization to resource (s) is called hacker h_i and represented as $H(h_i \in H)$ and $\forall H$(hackers) the access rights $a_{ih} \mapsto d_i = \varphi$ is true; a_{ih} is access rights of the hackers ($\mapsto$ implication to, and $\equiv$ is equivalent to). Using this information, we design two algorithms.

6.6.1 Algorithm 1:

If the query $Q(n_i, d_i)$ matches the n_i as owner for token identification (TokenID), then the corresponding utility function u_i will be generated, else the query reflects as $Q(n_i, hd_i)$, where h is a hacker.

If the hacker is an internal user then

$hu_i \supseteq u_i + h' d_i$ (u_i Internal user), alarms security manager about internal hacker.

   if $Q(n_i, d_i) \subset u_i$

else

if $Q(n_i, d_i) \not\subset u_i \&\& Q(n_i, d_i) \cong u_i + h' d_i$ then

Convert $Q(n_i, d_i)$ as $Q(n_i, hd_i)$ and generate $hu_i \supseteq u_i + h' d_i$

Store the user utility hu_i that contains $u_i + h' d_i$ and inform security and keep the counter (log) in alert for further attempts.

The Algorithm 1 helps to detect the hacker if the user tries to gain the information with unauthorized access from the database. The following query and Table 6.1 explains the unauthorized access to information.

$Q(n_i, d_i) \equiv Q(hn_i, d_i) \not\subset u_i$ or $Q(hn_i, d_i) \approx hu_i$ then

$Q(hn_i, d_i) = hu_i$ retrieve hu_i (utility from the Hacker alarm to database) and alert the security alarm, where hu_i is available in log or identified as a new hacker and logged as new entry. The log is provided in Table 6.1.

Table 6.1: Hacker Log and Action

Hacker	Status	Result	Action
A	new	hu_i	New hacker, alarm
A	repeat	hu_i	Alarm and freeze

In general, if the hacker attempts to gain access to the database at different times, the time attribute plays an important role to detect the hacker. The Algorithm 1 is modified as Algorithm 2.

6.6.2　Algorithm 2:

$Q(n_i, t_i, d_i)$ is genuine and attempted during duty times then corresponding utility function u_i will be generated,
else the query reflects as $Q(n_i, t_i, hd_i)$ then user will get $hu_i \supseteq u_i + h'd_i$ (where u_i Internal user information $h'd_i$ is the hacker alarm at time t_j).
if $Q(n_i, t_i, d_i) \subset u_i$ then exit(user access accepted)
else
if($Q(n_i, t_j, d_i) \not\subset u_i$&& $Q(n_i, t_j, d_i) \cong hu_i$)
Convert $Q(n_i, t_i, d_i)$ as $Q(n_i, t_i, hd_i)$ and generate $hu_i \supseteq u_i + h'd_i$ (alarm alert to Security manager)

Note: Store the user utility hu_i that contains $u_i + h'd_i$ and alert security and keep the counter for further attempts.

If the hacker is external then divert to the KDS. If the user hacks with authentication then the time stamp will help to detect the hacker. For example,
If $Q(n_i, t_j, d_i) \equiv Q(hn_i, t_j, d_i) \not\subset u_i$ or $\subseteq hu_i$ then $Q(hn_i, t_j, d_i) = hu_i$
Retrieve hu_i and alarm the security, where hu_i is available in log or identified as a new hacker and logged as a new entry. Table 6.2 provides the log entries.

Table 6.2: Hacker Log and Detection

Hacker	Status	Time	Result	Action
A	New, internal	Outside-bound	hui	Detect as internal hacker and alarm
A	Repeated, internal	Within-bounds	hui	Check for presence of real user and alarm and find real user

Depending upon the security level, the Algorithm 2 will be modified by adding the terminal type and log-on timings. Terminal type and time of access attributes along with access type attributes will protect the secret and top secret information.

Let us assume the hospital environment in the healthcare system. A doctor and nurse have same access rights to individual patient data (doctor prescribes the medicine and implemented by the nurse). Then the attributes patient id, type of medication, and scheduled time dose to be given to a patient are accessible by the nurse. The same attributes are also available by the doctor. Therefore, the system security depends upon the merge and decomposition of two or more users.

6.7 Conclusions

The issues and challenges in IoT [249],[36], [313], [404], [353],[60], [120], [331] processing of complex data in cloud and security issues were discussed in [391], [55], [193], [390]. We found that it is required to develop a trust and access control methodology in IoT and cloud environment for real-time access to data and processing for decision making. Therefore, in the current research, an objective function was proposed with a set of users, associated access rights, resources and return result verification. The proposed model is appropriate for the big data in a cloud environment where IoT devices are involved. Further, two algorithms were presented where the model can be extended to Hadoop distributed file systems to detect the external and internal hackers in a cloud environment. The tables were presented for hacker detection through algorithms. The user entry logs, authentication, and access rights have a significant role in providing the hacker information to the security administrator.

Chapter 7

A Layered Internet of Things (IoT) Security Framework: Attacks, Counter Measures and Challenges

Umang Garg

Department of CSE, GEU and GEHU, Dehradun, India

Preeti Mishra

Department of CSE, GEU, Dehradun, India

R.C. Joshi

Department of CSE, GEU, Dehradun, India

Internet of Things, also specify to as Internet of Everything (IoE), is a transforming buzzword due to the scope of the field and services. Smart applications have become a regular part of our day-to-day life with compact size and powerful features. The basic requirement of IoT devices such as sensors, actuators, Internet, and cloud storage can lead to various security issues in the IoT. To provide security features in IoT, there are four major concerns: device authentications, secure communication, protected information sharing, and secure storage unit. Several research bodies have been involved to find out probable security threats in IoT and provide countermeasure techniques to deal with them. However, there is no standard framework that can elaborate a layer by layer attack list in an organized way. The current chapter proposes a novel IoT security framework which discusses security issues and countermeasure techniques at various layers in IoT environment. Attack taxonomy has also been proposed and various attacks have been discussed concerning each layer. We also provide a real-time case study based on Denial of Service attack for home automation application. The state-of-the-art research challenges have been highlighted to provide future directions to researchers in this area.

7.1 Introduction

In the current scenario, the Internet of Things (IoT) is an emerging field that motivates researchers to design different applications. The basic definition of IoT is to connect objects and provide functionality anytime and anywhere whenever it is required by the end-user. Devices can interact with each other on the Internet with the help of unique address schemes known as Radio Frequency Identification (RFID). Electronic product code (EPC) is used to support RFID in the ubiquitous worldwide modern network [183] which provides standards to improve the visibility of objects. The term IoT was introduced by British technology pioneer Kevin Ashton in 1999, and in the year 2000 LG released the world's first Internet-enabled smart refrigerator [293].

There are several applications in the field of IoT such as smart city, smart agriculture, smart healthcare, smart home, smart industry, and so forth. Cisco is expecting more than 50 billion devices will be connected by 2020 with the world population 7.8 billion by 2020 [125]. In every second, 127 new IoT devices are connected to the Internet and it may surpass the mobile devices [21]. This expeditious growth of the traditional Internet into IoT has accelerated the countless domains of research which were unimaginable previously. Some IoT devices are used in our day-to-day life such as hospitals, homes, electronics appliances, and healthcare, etc. However developing a new application of IoT contains several challenges, i) object identification, ii) highly complex structure and distribution, iii) lack of common platform, iv) distinguishing protocols, and v) involvement of third party services [44]. Although there are various research challenges in IoT, this is a very innovative and progressive field for researchers.

The integration of physical devices (RFID, Sensors, Actuators, etc.) with the Internet and their communication can produce several challenges and security threats. Additionally, several key issues are involved with IoT such as mobility, heterogeneity, power-constraints, scalability, real-time actuation, and low energy devices, etc. [185]. IoT security has become one of the major issues since some of the high profile incidents have occurred due to a lack of security in IoT devices. Mendez et al. [261] explained an exhaustive survey regarding various security issues in different layers of IoT. Some traditional security issues were discussed by authors such as confidentiality, privacy, and availability [200]. However, traditional IT security issues are not sufficient for the IoT environment due to the diversity of hardware devices and multiple functional areas in IoT [158].

There are several security attacks that target IoT layers. In general, these attacks can be classified into five categories [56]: (i) Physical attack. These type of attacks can be possible with hardware devices when attacker tries to temper sensors or microcontroller (for example de-packaging, reprogramming the microcontroller, physically destroying, etc.). (ii) Side-channel attack. In this type of attack, an attacker can intercept the important information using some tools, for example, power analysis attacks, fault analysis attacks, EM attacks, etc. (iii)

Cryptanalysis attacks. These attacks are based on stealing the encrypted data or encryption key to obtain plaintext. For example, man-in-the-middle attack, ciphertext attack, etc. (iv) Network attacks. Wireless communication in IoT devices can generate several attacks during the transmission of information. Wireless transmission media is most vulnerable due to its broadcast nature [144]. For example, jamming attack, traffic analysis attack, false route, etc. (v) Software attacks. Software attacks are the source of different vulnerabilities in the system because of software support provided by the third-party vendor, for example, Trojan horse attack, worms, virus, etc.

To protect IoT applications and devices, there are several techniques to provide security solutions from different attacks of IoT. Akayev et al. [29] contributed to the area of information security and evidence identification in IoT network. They proposed a real-world scenario user-centric IoT (UCIoT) and risk assessment. The low cost and diverse nature of IoT devices are vulnerable to the physical, side channel, and clone attacks. Physical unclonable functions provide security solutions from these types of attacks [43]. Pajouh et al. [295] presented a model for intrusion detection based on two-layer dimension reduction and a two-tier classification module. These modules are designed to detect intrusive activities in IoT networks, specifically low-frequency attacks like U2R and R2L. Some authors surveyed IoT security from the perspective of IoT architecture. Lin et al. [227] investigated IoT security challenges at each layer in three-tier and four-tier architecture. Authors also provided a relationship between IoT and fog computing with their challenges and security issues.

Minoli et al. [269] discussed the Blockchain mechanism to secure IoT oriented applications in the context of the defense-in-depth approach. According to the authors, Blockchain is a combined security mechanism including encryption, a secure operating system, and a trusted party environment. Mukherjee et al. [282] addressed IoT based security requirements and provided a flexible IoT security middleware for end-to-end cloud communication. Authors also implemented test beds to speed up secure communication. Pablo et al.[303] presented CoAP based framework to provide access control for low power devices. According to authors CoAP protocol is one of the best suitable protocols for low bandwidth links like 6lowpan over IEEE 802.15.4.

In this chapter, we provide a detailed survey of IoT attacks and propose a security framework for IoT enabling cloud storage services. We investigate several research challenges in IoT such as heterogeneity, scalability, and interoperability. We demonstrate a real-time experimental setup to perform a DoS attack. It can de-authenticate the home automation gateway to stop the required facilities of the system using an unauthorized way. The contributions of this chapter are summarized as follows:

- We propose a novel security framework for IoT applications that can provide a safe environment for secure communication.

- We propose taxonomy on IoT attacks that target different layers of the security framework.

- We provide state-of-the-art research challenges in IoT security.

- We also demonstrate a real-time attack scenario of denial of service attack as a case study on home automation application.

The rest of the chapter is organized as follows: Section 7.2 discusses the summary of the related work of some researchers in the field of IoT security. Section 7.3 contains a classification of possible attacks in an organized manner. We propose a novel security framework with a description of vulnerabilities and attacks concerning each layer in section 7.4. Section 7.5 defines the case study of practical implementation of Denial of service attack in home automation application. Finally, section 7.6 and 7.7 discuss several research challenges and conclusion of the chapter.

7.2 Related Work

Internet of things (IoT) is an emerging field which motivates the researcher to design different applications in several domains. The IoT refers to three key elements: sensors (used to sense data from the physical environment), network (used to send data from sensor to storage unit), and actuator (used to perform actions). Rather than these services, it also includes some services such as cloud services, machine learning algorithms, number of users, and some analytical servers. Each service requires some foundational ethical issues like privacy, information security, physical safety, and trust [37]. There are several architectures proposed by researchers; here we discuss one or two of them. Authors [54], introduced a fundamental three-layer architecture: the sensing layer, network layer, and application layer. The IoT has added a new dimension to the Internet and enables communication with hardware devices leading to create smart objects which can be accessed anytime and anywhere. However, there are several issues in IoT such as heterogeneity, scalability, and security, etc.

Suo et al. [348] proposed a secured architecture based on four layers: perceptual layer, network layer, support layer, and application layer. Some challenges are discussed by authors for each layer as well as different aspects of information security. Bandyopadhyay et al. [58] proposed a five-layered architecture that can provide a standard framework for industries. The edge technology layer and access gateway layer can be used for data collection and data handling. The Internet layer provides communication between two bottom hardware layers and the top two application layers and it is very critical to handle that operates in the bidirectional mode. Top two layers (middleware and application) are responsible for the delivery of several applications which belong to the different domains.

Atzori et al. [54] defined the different visions of IoT with their enabling technologies that can help to understand IoT at the primary level. Lin et al. [227] discussed a comprehensive review of IoT including architecture, and the integration of enabling technology such as fog or edge computing in IoT applications. The authors discussed some security and privacy issues that could affect the performance of applications with their solutions. Al-Fuqaha et al. [31] presented an overview of IoT with horizontal integration between application layer protocols and different aspects of IoT. A few challenges and issues related to the design and development of IoT applications is presented by several authors. This paper also defines the role of cloud and fog resources in the delivery of IoT resources. Some detailed application use cases were also presented to integrate IoT devices with protocols.

Andrea et al. [47] explored the security goals required in a secure IoT environment. They categorized IoT attacks in four classes such as physical, network, software and encryption attacks. Zhou et al. [403] addressed the privacy and security challenges of cloud-based IoT. They also proposed a new method for preserving data aggregation without exploiting public key encryption. Serrano et al. [162] addressed privacy issues and rate privacy risk in the IoT domain. To provide rating to the risks, open web application security project (OWASP) RRM methodology was adopted by the authors. Lin et al. [226] surveyed some privacy techniques in various applications of IoT. Some security requirements for smart home were also discussed as well as suggested suitable security architecture for IoT. Ziegedorf et al. [405] provide a detailed analysis of threats related to privacy in IoT. Threats are classified into seven categories such as identification, localization, profiling, privacy violation interaction, life-cycle transition, inventory attack, and linkage.

Dawoud et al. [107] proposed a secure framework for IoT based on Software Defined Network (SDN). They focused on IoT deployment where security is a critical factor. The authors discussed improvement in SDN-based IoT architecture and deployed an intrusion detection system based on machine learning. Martino et al. [113] compared and analyzed different architectures of IoT based on industrial and academic research. In this paper, researchers also identified a real-time case study and described the most common configuration of IoT devices. The authors discovered that there are two main research challenges in IoT, IoT security and interoperability.

Chaqfeh et al. [87] investigated some major challenges and solutions in the middleware layer. The authors classify the solutions in three categories: semantic web and services, RFID sensor network, and robotics-based system. The authors addressed middleware challenges except for the scalability of the service. The semantic web solution is one of the best solutions to address service discovery and service composition in the IoT area. Teixeira et al. [227] investigated about the challenges associated with service-oriented IoT middleware. They proposed a viable technique for service discovery and service composition with the help of a probabilistic approach to handle the problems of scale, unknown availability of service and accuracy of data.

7.3   Taxonomy of IoT Attacks

IoT may have significant economic and social benefits in society. However, privacy and security are the major issues that are remaining in IoT applications. There are various applications in which security and privacy are often neglected or come in to the manufacturer's mind as a second thought. This is due to market competition, and cost deduction during the developments of the applications. Some potential attackers take this as an opportunity to take unauthorized access to the devices, data or the whole system. Security requirements can be mainly classified into three categories CIA-triad. Confidentiality provides limited access to an unauthorized users to access limited information. Integrity is the requirement of reliable services between sources and destinations. IoT devices are more vulnerable compared to IT security techniques because of the availability of IoT devices in an open environment. We can classify IoT attacks in four categories [159]. Various attacks can target different layers in IoT as shown in Figure 7.1.

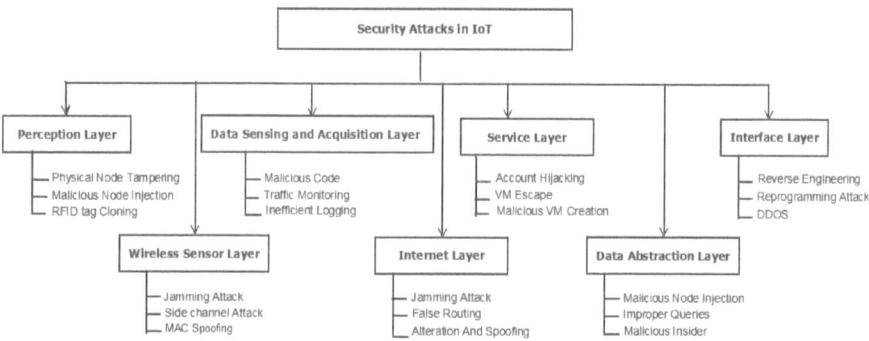

Figure 7.1: Attack Classification of IoT

7.3.1   Physical Layer Attacks (PLA)

The main objective of this layer is to collect information from the physical environment and transmit the information after converting it into digital signals. This layer is most vulnerable due to easy availability of the IoT devices such as sensors, actuators, RFID, micro-operating systems. The information provided by these sensors can be about the location, motion, temperature, light, etc. The information collected from sensors is shared in a local network like Zigbee, or the Bluetooth based network. Several attacks can be triggered at this layer; some of them are:

7.3.1.1 Physical Node Tampering

In node tampering attack, devices could misbehave or even destroy the whole system, which is a big challenge for applications. This type of attack can be possible due to the availability and accessibility of physical devices so that an attacker could manipulate the circuit, glitch the clock, modify in the tag, or physically destroy the sensors [39]. For example, if a setup is created for a fire tracking system and placed somewhere in the forest to detect the fire, an attacker could get access to all sensors and manipulate the functionality of the system or circuitry board.

7.3.1.2 Malicious Node Injection

In this attack, an attacker can inject more than one malicious nodes in the existing system so that it can pass or manipulate the data to the authorized nodes. The objective of this attack is to have unauthorized access to the network or control the other devices accordingly. A malicious node can prevent the successful delivery of the original message and send a false message to the network [109]. This attack is also known as a man-in-the-middle attack. To detect the injection of a malicious node, a MOVE (Monitoring Verification) technique can be used to identify malicious behavior of nodes and decide whether a node is malicious or not.

7.3.1.3 RFID Tag Cloning

In this attack, an attacker creates a duplicate identity of the existing tags so that a false user can be treated as an authorized one who can access all data or manipulate the information. Some of the examples of tag cloning are bank ATM cloning, identity cards to access restricted areas, and confidential information. Each RFID tag has its unique EPC (Electronic Product Code) that is provided at the time of integration of the RFID tags by EPC global network [186]. A successful tag cloning may lead to several attacks, financial losses, or serious problems for commercial applications. Although RFID tags are based on cryptography and encryption may be able to prevent delay, some anti-cloning mechanisms are required, however, to support tag cloning detection.

7.3.2 Wireless Sensor Network Layer Attacks (NLA)

The flow of information among devices at the first layer and the third layer can be possible with the help of a wireless sensor network layer. Wireless network devices can communicate using wireless networks such as IEEE 802.15.4, Wi-Fi, BLE, LoRaWAN, and LTE. Several attacks can occur at this layer; some of them are:

7.3.2.1 Jamming Attack

In the perception node layer, radio signals can be jammed with a Radio Frequency transmitter. Jamming attack can be classified into three categories: i) Constant Jamming: In it attacker transmits continuous random bits so that readers can deny its services. ii) Deceptive Jamming: An attacker can be sent a continuous stream of packets to create abnormal operation of the system. iii) Random Jamming: In this attack, attacker quickly sends jamming signals to the devices. All these jamming attacks can be controlled using regulated transmitted power, and direct sequence spread spectrum [123].

7.3.2.2 Side Channel Attack

In this type of attack, an attacker can intercept important information using some tools. For instance, Nia et al. [289] described the side-channel attack which is based on electromagnetic (EM) radiation, released by an object which may have important information. Electromagnetic radiations can be classified into two categories: i) unintentionally generated electronic component can emit EM waves that may be used for side-channel information; ii) intentionally generated medical components that can use EM waves to transmit some data wirelessly [354]. The EM wave can be detected using some spectral analyzers that require static carrier signal of static amplitude. So, the unintentionally generated EM signals can remove demodulation.

7.3.2.3 MAC Spoofing

Whenever a personal area network is formed, a malicious attacker can spoof a MAC address during the encrypted key generation. Attackers can spoof MAC addresses that can disconnect legitimate users or modify information during transmission. There is no policy to prevent this attack; however, we can take long variables, special characters, and numbers for the pairing of the devices [266].

7.3.3 Data Sensing and Acquisition Layer Attacks (DSAL)

This layer provides cloud-like environment at the network edge that can filter data before moving to the Internet [104]. It can handle data explosion conditions that can occur on the Internet. This layer tries to save channel bandwidth because of the removal of ambiguity and duplicity in data. Although this layer is less vulnerable, there are some attacks can target the functionality of this layer.

7.3.3.1 Malicious Code

There is no sufficient validation scheme of the input in data acquisition. In such a case, an attacker inserts some malicious code or injects it to the service provider and then the desired action must be performed based on instructions.

A hardware component may be attached at the lower layer (data sensing and acquisition layer) to insert some malicious code which either tries to access user data or executes instructions to non-validate the process [74]. Pre-testing is a mechanism that can be helpful to handle these kinds of attacks.

7.3.3.2   Traffic Monitoring

The information collected from sensors and hardware components can be monitored by an attacker with a false identity or false node. Sensed data have common patterns in the data or sequence of similar events that can be aggregated to find out the information using some probability cases and patterns [164].

7.3.3.3   Inefficient Logging

To detect a hacking attempt by an unauthenticated user, logging is a mechanism that provides log events for unsuccessful attempts or application errors. If there are more unsuccessful attempts within a time frame, services of the system can be stopped. To encrypt the log files, we can prevent from inefficient logging detail [145].

7.3.4   Internet Layer Attacks (ILA)

Network layer is mainly responsible for connectivity among all devices and communication between hardware and the cloud server or end-user. This layer aggregates the data from different devices and provides routes for a specific device or the user through a gateway. This layer is vulnerable due to the global scope of the data; so several attacks can be possible at this layer.

7.3.4.1   Jamming Attack

Jammers of this layer are energy inefficient when compared to physical layer jamming attacks. In this attack, attackers focus to jam data packets and ACK messages as well [123]. Jamming of data packets depends on the type of MAC protocol used in communication between nodes, in which the attacker tried to manipulate some bits of packets by interfering with communication. It is one of the fatal attacks which can block the channel by generating false packets to introduce noise in the channel. IoT is a field in which all physical devices have limited energy or power constrained so the jamming attacks can drain these resources. Regulated transmitted power and frequency hopping spread spectrum are the countermeasures for jamming attacks.

7.3.4.2   False Routing

An attacker tries to generate or transit false routing information to the nodes connected in the network. False routing can damage the packets or leak the information transmitted over the false link. Four scenarios can generate a false route: i) false route error message- if network protocols do not have any route up to destination node then it sends a route error message to the source and the link is broken. Every time this error message can truncate the communication among nodes. ii) Poisoning route-cache- If any packet contains route information in their header update route cache, it can exploit by suspicious node and send a spoofed packet with manipulated route information to mislead the packets. iii) Overflow routing table- A malicious node can generate a false node with overflow of routing information for non-existence paths. iv) Rushing attack- it is like a sink-hole attack that can absorb all packets of the network with false route information and control over the network with its modification [187].

7.3.4.3 Alteration and Spoofing

In a routing protocol, each node has its rank that increases from root to child. An attacker can modify the rank of any node to attract child node and network traffic towards the root node. Due to this attack, routes may not be optimized or a loop is created in the route that can detect with version number and rank authentication mechanism and Trust Anchor interconnection loop [306].

7.3.5 Service Layer Attacks (SLA)

The responsibility of the service layer is very important due to interfacing between network data and the application. An application interface, web service, cloud storage, and data centers are the major components at this layer. These services are provided by third-party vendors; that's why these are the most vulnerable parts of the IoT applications. Although the service layer is provided by reliable sources, it has several security flaws and attacks. Some possible attacks are as follows:

7.3.5.1 Account Hijacking

Account Hijacking is one of the biggest challenges in cloud services. Several attacking mechanisms are used to access credentials of the users such as phishing with the password. These attacks take benefit of software vulnerability or clone identity. If an attacker can access credentials then it may harm the information, manipulate data, or can eavesdrop on the important information. A weak password, insufficient authorization, and inefficient input validation schemes are the main reasons to generate this attack. In June 2014, Amazon AWS failed to protect the administrative interface with an authentication scheme [291]. Dynamic credentials and access management guidance are two countermeasure techniques that can be used to prevent Account Hijacking.

7.3.5.2 VM Escape

Virtual machine programs (VM) can analyze the behavior of run-time data dynamically. So to detect any modern attack it requires VM memory and VM monitor [272]. In this attack, the attacker can access the memory which is beyond the access of tenant VM. An attacker can breach the isolation of VM and can manipulate other VMs. The major objective of this attack is to configure flexibility, and code complexity. Confidentiality, data integrity, and privacy are the major concerns of this attack. Trusted cloud computing and virtual datacenter are countermeasures techniques to handle VM escape [271].

7.3.5.3 Malicious VM Creation

An attacker could create a legitimate VM account that may have malicious code injected in a normal program that works as a self-explanatory code [271]. In this attack, the attacker can destroy some system files, user data, or damage the whole system by replicate viruses and worms. To construct a secure and high-performance network, Mirage is a single kernel cloud computing platform to deploy cloud services through applications.

7.3.6   Data Abstraction Layer Attacks (DALA)

In IoT applications, data collected from several devices can be transmitted further; it may lead to the data explosion. Normalization, Consolidation, or indexing are the main techniques to improve data quality and network performance for further analysis of stored data. To improve the overall performance of the application, we require a faster response from cloud or data servers; the data abstraction layer is the key layer to provide this functionality. Amazon IoT, Amazon Green-grass, Dell-Statistica, and Azure are some analytics tools that extract the data in real-time scenarios [406]. Although, this layer is less vulnerable, some attacks can be possible at this layer.

7.3.6.1 Malicious node Injection

This is one of the most common attacks which can occur at this layer, in which attacker can insert some malicious code in the form of a string that is sent to the SQL server for malfunctioning of the application. If any system does not have sufficient code checks, it may attract the attackers and inject some malicious code to misuse or disrupt the application. Cross side scripting can be used to inject the code and hijack the account of the user. Firewall or security checks are the countermeasure techniques for malicious code injection [109].

7.3.6.2 Improper Queries

In this attack, the attacker wants to gather possible information about the structure of the table and fields of the table. The attacker may generate some error message to gain access on the behalf of a legitimate user and gain full access to data. Some error messages which are received from the database can guide the attacker. After getting proper guidance, an attacker can damage the system or misuse it. Some predefined statements like PREPARE supported by many databases provide a template for SQL queries [74].

Table 7.1: Attack Taxonomy

Layer	Attack	Behavior	Target	Countermeasures
PLA	Physical Node Tampering [39]	Can Modify or destroy the system	Hardware Components	Tampering proof design
	Malicious Node Injection[109]	A false node can be placed in system	Prevent original information	Monitoring Mechanism, IDS
	RFID Tag Cloning [186]	Access data in an unauthorized manner	False Information	Electronic Product Code
NLA	Jamming Attack [123]	Jammed Radio Signals	Communication between components	Regulate transmitted power or DSSS
	Side-Channel attack [289]	Eavesdropping	Sensitive information	Spectral analyzer and modulation
	Mac Spoofing Attack[266]	Malicious attacker can spoof the MAC	Disconnect the legitimate user	Long and variable password
DSAL	Malicious Code [74]	No sufficient Validation Scheme	Affect to the Service Provider	Pre-Testing Mechanism
	Traffic Monitoring [164]	Performance monitoring by false node	Data Leakage	Probability cases and find the pattern
	Inefficient Logging [145]	Detect log based events	Data pattern on the basis of log details	Encrypt the Log File
ILA	Jamming Attack [123]	Jammed Radio Signals	Communication between components	Regulate transmitted power or DSSS
	False Routing [187]	False Routing Information to mislead	Damage the Packets	Bi-Verification of the Route
	Spoofing Attack [306]	Modify the rank of node	Attract child node towards the root node	Trust Anchor interconnection Loop
SLA	Account Hijacking [47]	Phishing with password	Authenticated account	Dynamic access management
	VM Escape [272] [271]	Breach the isolation of VM	Manipulation of VM	Trusted cloud computing
	Malicious VM Creation [271]	Legitimate accounts have malicious code	Damage some system files	Mirage
DALA	Malicious code Injection [159]	Some SQL Injections	Malfunctioning of the application	Firewall or Security checks
	Improper Queries [289]	Gathering Information of table structure	Gain full access on data	Prepare statement
	Malicious Insider [271]	Former employees of the company	Confidentiality or integrity of data	Auditable process or logging
ILA	Reverse Engineering [297]	Analyze the software	Gain sensitive data	Tamper proof software
	Reprogramming attack [364]	Modify code from remote site	Misbehavior of the system	Secure Programming process
	DDOS [48]	Continuous overwhelming of packets	Stop the service	Internet Firewall

7.3.6.3 Malicious Insider

A malicious insider is a threat in which a current or former employee of the organization, having authorized access to the data, misuses or shares the data with some third party intentionally for personal benefits. It can affect the system's confidentiality and integrity of data or information. Malicious insiders are difficult to detect due to their authenticity and full accessibility of services. Cloud service provider's key management is different from the data storage unit in an encrypted way so that unauthorized access can be prevented. Auditable process, effective logging, Segregate departments are some countermeasures of this attack [159].

7.3.7 Interface Layer Attacks (ILA)

Some software or application programs are incorporated with cloud servers or APIs which provide an interface to the end-users. There is no common standard for this layer due to the heterogeneous behavior of applications. The security issues are different according to the application. There are two major issues of this layer, data theft and privacy. Additionally, some attacks are:

7.3.7.1 Reverse Engineering

In the application of IoT, the attacker can analyze the software to gain sensitive information or some credentials of users. With the help of reverse engineering, the attacker can use the vulnerability of the programming errors and can leak or exploit the software or IoT objects[297]. Tamper-proof software can prevent reverse engineering.

7.3.7.2   Reprogramming Attack

If an attacker reprograms any IoT object from the remote site using a network programming system then it may misbehave from its normal functionality. If there is insufficient protection at the programming process, the attacker can modify all its functionality and control some parts of the application. This is the most dangerous attack at this layer because it can attack privacy, integrity, confidentiality and much more. So we have to apply a secure programming process to prevent reprogramming attacks [364].

7.3.7.3   DDoS Attack

Distributed Denial of Services is the attack in which an attacker temporarily instructs the number of Internet-enabled devices known as IoT botnet and then sends continuous requests or packets to the server to access its services, so it may overwhelm the server and stop its proper functionality. A DDoS attack can exhaust the channel bandwidth or jam the server of IoT objects. DDOS attacks are classified in two categories: i) reflection in this attack, (the attacker sends packets with false IP address); ii) amplification (a large number of packets can overwhelm the server). Internet firewall can periodically monitor the suspicious traffic to prevent DDoS attack [48].

An attack taxonomy depicted in Table 7.1 that contains attack behavior, its target device, and countermeasure techniques can be applied to deal with the particular attack.

7.4   Proposed IoT Security Framework

Several architectures were proposed by different authors and researchers [58], [44], [31]. All have several layers from the sensing layer to the application layer according to the requirements of industries, applications, and societies. A three-layer reference model [23] was the first reference model in which authors described an extended version of a wireless sensor network with cloud services. Another four-layer model [348] is an alternative that has been proposed to interact with a complex system. And a five-layer reference model [58] was a generic layered architecture for IoT based on service-oriented architecture (SOA). This model has the potential to interact with several applications and well-defined components. It is observed there is no standard framework that can provide a security mechanism or a well-recognized manner. Hence, there is a critical requirement of a framework that can provide an operational guarantee for IoT applications to bridge that gap between physical devices and the virtual platforms. We propose a security framework for IoT as shown in Figure 7.2. It consists of seven layers. The description of every layer is as follows:

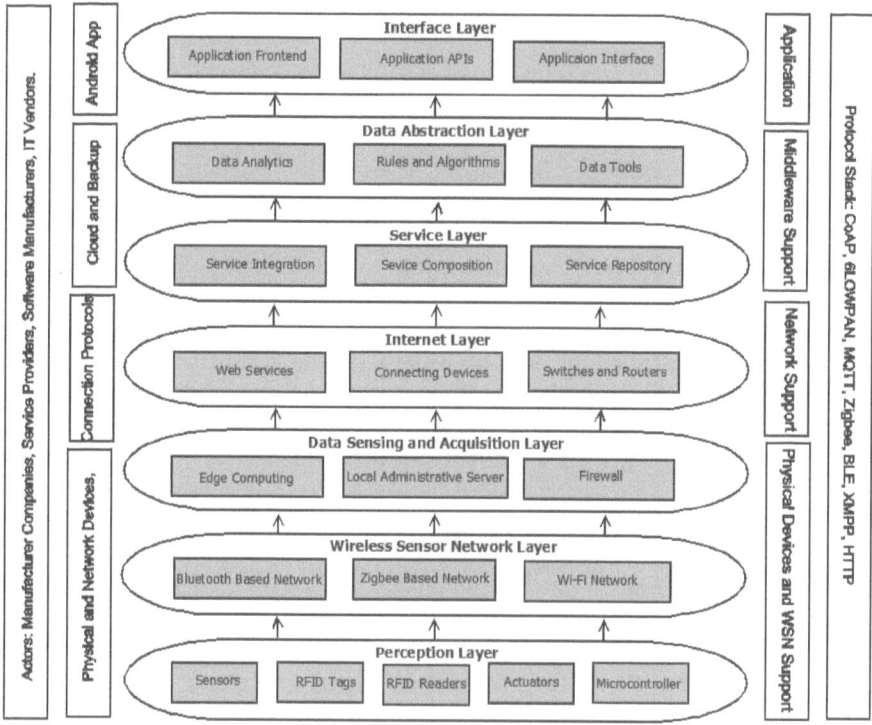

Figure 7.2: Security Framework for IoT

7.4.1 Perception Layer

This is the lowest layer that consists of several physical devices like sensors, actuators, Microcontrollers, RFID tags, embedded systems, micro-operating systems, and RFID readers, etc. Data are captured or sensed by different sensors and shared between the components by using a network. The main concerns of this layer are the deployment of the nodes, heterogeneity of devices, Cost, size, and energy consumption by end nodes. The Perception layer is the most vulnerable layer in the security framework because of the availability and accessibility of the hardware components. Some natural calamities (such as earthquakes, floods, storms, etc.) and environmental threats (such as fire, chemical accident, etc.) can destroy the whole infrastructure of the system. An attacker can easily deploy an attack to a sensory node and can also modify the data collected from the sensors. There are a wide variety of security concerns [154] at this layer including device authentication, trusted devices, physical protection, and temper proof designing [34]. Hash-based techniques, intrusion detection systems and granular segmentation are some of the techniques to deal with several attacks.

7.4.2   Wireless Sensor Network Layer

Wireless sensor network has enabled the low cost and low power network that can collect information from heterogeneous sensors. In this network, there are two components: aggregation and base station. Aggregation point is used to collect information from nearby sensors. Then the information is integrated and sent to the base station to process the collected data [274]. Various types of networks exist at this layer such as Bluetooth, Zigbee network, Infrared network, and Wi-Fi network. This layer is vulnerable due to unencrypted information movement and unprotected communication channels. Some of the different attacks that can be possible with wireless sensor network are injecting false data in WSN, Impersonating, unauthorized access, overloading the WSN, and monitoring and eavesdropping [366]. To deal with different attacks, some of the techniques can be used such as limiting the administration control rate, access control, secure routing, and strong and proper authentication techniques.

7.4.3   Data Sensing and acquisition layer

Sensors and hardware devices can collect ample data that are not useful for further processing and analysis purposes. Hence there is a requirement of data abstraction to get a faster response in real-time applications. This layer is used to collect data on a local server or a gateway to process and extract useful data. The main features of this layer are the collection and filtering of data, triggering the event, data aggregation, and gateway to the network. Some vulnerabilities of this layer include insufficient validation, inadequate testing mechanism, and information leakage [20]. Malicious code, traffic monitoring, and inefficient logging are major attacks that can target this layer. To deal with these kinds of attacks different techniques are available, such as a pretesting mechanism, encrypting the log file, and pattern finding.

7.4.4   Internet Layer

This layer is also known as a communication layer or network layer. The data abstracted in previous layers can be shared in remote places or far from the physical setup using the Internet. The Internet can have billions of interconnected devices that use traditional Internet protocol (TCP/IP). A large range of information and data services are provided by the Internet like the connection between email sharing, World Wide Web applications, etc. Internet layer is responsible for the routing of the packets, Plug-ins, the protocols, IP based communication, network security implementation, and reliable delivery of packets. Some of the vulnerabilities at this layer are: IP address spoofing, route spoofing, wireless access points, and vulnerable transmission media. There are several attacks at this layer such as low rate denial of service, traffic analysis attack, false routing, eavesdropping, and spoofing [309]. To deal with different attacks, some techniques are Hilbert Huang transform, tools to analyze packets, bi-verification of the route, deterministic path loss model, and trust anchor interconnection loop.

7.4.5   Service Layer

The service layer is a kind of middleware that is an enabler of services and applications. This layer is designed to provide a common platform to the applications of IoT with common application programming interfaces (APIs) and protocols [64]. The main responsibilities of this layer include information storage, data processing, analytics services, integration of services, and event processing, etc. To enable any service with an application of IoT, the following components are used: service discovery, service composition, and service APIs, and trust management [222]. Some main actors of this layer are cloud services, backend services, database and storage management, and data storage components. Some vulnerabilities of this layer are the reliability of service, insecure cryptography, data protection, and Internet dependency. The security requirement of this layer includes authorization, service authentication, privacy protection, anti-replay, and availability. Data loss and modifications, VM escape, malicious VM creation, insecure VM migrations, and brute force attacks are some attacks that may occur at this layer [271]. To handle these attacks, some of the techniques can be applied such as backup and retention, trusted cloud computing, mirage, VNSS, and site scanner [248].

7.4.6   Data Abstraction Layer

For better performance of any application, it is required to relinquish some data and enhance data storage. Sensors can generate repeated data that, at the same time, can lead to the delicacy of data that cannot be handled at the application layer [277]. So the normalization, consolidation, filtering, and indexing are the ways through which data can be controlled for further analysis. Some rules and algorithms, decision-making analyzers, and big data tools can simplify the data. Some responsibilities of this layer include reformatting of data, preserving data for an authentic user, normalizing and indexing data for faster response [227]. Software vulnerability, redundant data, and sensitive information leakages are security vulnerabilities that may occur at this layer. Some attacks that may occur in this layer are an excessive privilege, improper queries, and the malicious insider. To handle such attacks, authentication mechanisms, access control policies, preparing statements, auditable processes, and effective logging are major techniques.

7.4.7   Interface Layer

This layer includes several interfaces for a variety of applications from small RFID applications to large, smart city applications, which can be implemented using standard protocols. This is the highest layer at which users and different computing devices can interact with the smart system using cloud services or application software [34]. This layer provides information interpretation with the help of software cooperation between the cloud server and its applications. There are different actors to support at this layer like analytics and visualization tools, IoT support applications, web sites, and cloud software, etc. Some security vulnerabilities like third party failure, software bugs, unauthenticated access, and configuration errors may generate serious issues at this layer. The attacks of this layer include malicious code injection, reprogramming attack, DDoS, reverse engineering, backdoor, and phishing attack [144]. Some security mechanisms that can be applied to handle those attacks are security checks, internet firewall, temper proof design, lightweight cryptography algorithms, etc.

7.5   Case Study: Implementation of Denial of Service Attack in Home Automation

Recently, many literatures have been published in the field of IoT security. Some of them were dealing with privacy, authentication of a user, trust management, etc. And another group of researchers works with the several attacks made possible in IoT. We are dealing with one of the attacks known as Denial of Service attack.

7.5.1   A brief description of attack

As the number of application areas increases using IoT, the vulnerability of the system will increase. Although IoT provides substantial benefits to the users, there are various security challenges implicit with the system. Denial of Service (DoS) [393] is one of the major attacks, in which the attacker attempted to prevent an authentic user to access the services. In DoS attack generally, attacker floods the data on the network to block the channel to prevent the access of other legitimate users. The attacker sends the messages to the server and asks the server to authenticate the request with a false return address. The server does not detect the false address of the attacker, leading to the waiting state till the termination of a connection. When the connection is closed, the attacker again sends more messages with a false address. The server starts the authentication process again; this procedure is repeated, leading to the waiting state of legitimate users [298]. DoS attacks can exploit security vulnerabilities in the network or the server. Some DoS attacks were implemented in history: i) Smurf. In this, an attacker used the broadcast address of the network by sending some spoofed packets and flooded the targeted IP address; ii) Ping flood. In this type of attack, attacker floods ping packets to overwhelm the target server; iii) Ping of death. In this type of attack, a malformed packet is sent to the target machine that can crash the whole network or server.

7.5.2   Experimental Test-bed Details

To implement the Denial of Service attack, we setup a home automation circuitry in which a Raspberry Pi 3B+ module is used to control lights, fans, and other home appliances. To implement this setup, we have to install Raspbian or Noobs 3.2.0 operating system using some external storage such as a micro storage card. After installation of the operating system, Raspberry Pi can coordinate with other devices using some small code upload in the system. After the configuration of the system, we can control the home devices like fans, lights, and other devices using the Internet from a remote place. Now, Raspberry Pi provides facilities using the Internet or home gateway. In case an attacker wants to access this gateway in an unauthenticated way then it has to hack the home gateway. In this scenario, we have performed DoS (distributed denial of service) attack on the home gateway to de-authenticate the raspberry Pi from the network. We will send de-authenticated packets to the gateway in a large amount. To perform this, we have to install KALI LINUX 2019.1 operating system with 4.19.13 kernel version for this experimental setup. And some tools such as airmon-ng, AIRODUMP-NG, AIREPLAY-NG are used to perform monitoring and for accessing purposes. (These tools are inbuilt with the kali Linux.) To perform the DoS attack, we have to follow some steps:

7.5.3   Execution Steps

Step 1- Putting Wi-Fi adapter in monitor mode using airmon-ng tool First of all, we have to enable our network interface card in monitor mode. To check the functionalities of an interface card, the command is airmon-ng start wlan0mon as shown in Figure 7.3. In this command airmon-ng is used, which is a tool of Kali Linux and wlan0mon is an interface card (change according to the machine). It provides details about all running processes in the background.

```
root@kali:~# airmon-ng start wlan0mon
```

Figure 7.3: Monitoring of Wi-Fi Adapter

Step 2- Abort running processes After the execution of the above command, we have a list of running processes. Each process can be identified with its process id and process name. These processes must be killed by using a command kill process-id so that there is no interruption in the background. In the current case, the terminal shows three processes with id 1252, 1308, and 1344 running on an interface and a chipset as shown in Figure 7.4.

Step 3- Capture the traffic When all background processes are killed, we can capture the wireless traffic that lies in our Wi-Fi range. Now with the help of Airodump-Ng tool, wireless adapter can be set in capturing mode using a simple command Airodump-Ng wlan0 as shown in Figure 7.5. This command is used

```
root@kali:~# airmon-ng start wlan0mon

Found 3 processes that could cause trouble.
If airodump-ng, aireplay-ng or airtun-ng stops working after
a short period of time, you may want to run 'airmon-ng check kill'

  PID Name
 1252 NetworkManager
 1308 wpa_supplicant
 1344 dhclient

PHY     Interface       Driver          Chipset

phy0    wlan0           rtl8723be       Realtek Semiconductor Co., Ltd. RTL8723B
E PCIe Wireless Network Adapter

root@kali:~# kill 1252 1308 1344
```

Figure 7.4: Abort Running Process

to detect all MAC addresses of devices lies in a particular range. This command

```
root@kali:~# airodump-ng wlan0
```

Figure 7.5: Capture Traffic

provides monitoring of all wireless devices that facilitate nearby and the output (as shown in Figure 7.6) will be like- where BSSID is the MAC address of the gateway, and PWR provides the information about station number and data rate with authentication technique.

```
CH 14 ][ Elapsed: 0 s ][ 2019-07-18 16:14

BSSID               PWR  Beacons    #Data, #/s  CH  MB    ENC   CIPHER AUTH ESSID

96:14:7A:10:08:B4   -61        2         0    0  13  54e.  WPA2  CCMP   PSK  vivo

BSSID               STATION           PWR   Rate   Lost    Frames  Probe
```

Figure 7.6: Output window with MAC address of Gateway devices

Step 4- Focus on target access point In the above output, we have different gateways with their BSS ID and channel number. Thus, we can check every gateway device one by one if we do not have the MAC address of the target device. To detect the MAC address of the targeted device, which facilitates the home gateway, open the terminal again and type command: airodump-ng –BSS ID 96:14:7A:10:08:B4 -c 7 wlan0 (as shown in Figure 7.7). Where BSS ID is MAC address of the access point and c is the channel number. The output of the above

```
root@kali:~# airodump-ng --bssid 96:14:7A:10:08:B4 --channel 7 wlan0
```

Figure 7.7: Focus on Target Access Point

command (as shown in Figure 7.8) contains information about the particular access point with its MAC address, Beacon number, rate, frame number, etc.

```
CH  7 ][ Elapsed: 42 s ][ 2019-07-18 06:54 ][ fixed channel wlan0: 11

BSSID              PWR RXQ  Beacons    #Data, #/s  CH  MB    ENC  CIPHER AUTH E

96:14:7A:10:08:B4  -73  19      124        0   0   7  54e. WPA2 CCMP   PSK  v

BSSID              STATION            PWR   Rate    Lost   Frames  Probe

96:14:7A:10:08:B4  B8:27:EB:72:AC:F8  -63   0 -24e    0       5
```

Figure 7.8: Output window with MAC address of Access Point

Step 5- Perform the attack We have the MAC address of Raspberry Pi under STATION section. Now, we can flood the data packet to the network to JAM the traffic. Or we can de-authenticate the gateway by sending a large amount of traffic on the home gateway leading to perform DoS (Denial of service) attack. In our example, we will use MAC address of Raspberry Pi and MAC address of access point which is B8:27:EB:72:AC:F8 and 96:14:7A:10:08:B4, respectively. Now, type a small command to perform the attack: Aireplay-ng -0 0 -a 96:14:7A:10:08:B4 -c B8:27:EB:72:AC:F8 wlan0 as shown in Figure 7.9. Where -0 is a de-authentication attack and 0 is the number of packets sent to the access point. It may vary 100 or 200, etc.

```
root@kali:~# aireplay-ng -0 0 -c B8:27:EB:72:AC:F8 -a 96:14:7A:10:08:B4 wlan0
06:59:06  Waiting for beacon frame (BSSID: 96:14:7A:10:08:B4) on channel 7
06:59:07  Sending 64 directed DeAuth. STMAC: [B8:27:EB:72:AC:F8] [20|24 ACKs]
06:59:07  Sending 64 directed DeAuth. STMAC: [B8:27:EB:72:AC:F8] [65|94 ACKs]
06:59:08  Sending 64 directed DeAuth. STMAC: [B8:27:EB:72:AC:F8] [ 6| 8 ACKs]
06:59:09  Sending 64 directed DeAuth. STMAC: [B8:27:EB:72:AC:F8] [74|120 ACKs]
06:59:09  Sending 64 directed DeAuth. STMAC: [B8:27:EB:72:AC:F8] [115|216 ACKs]
06:59:10  Sending 64 directed DeAuth. STMAC: [B8:27:EB:72:AC:F8] [33|61 ACKs]
06:59:10  Sending 64 directed DeAuth. STMAC: [B8:27:EB:72:AC:F8] [17|35 ACKs]
06:59:11  Sending 64 directed DeAuth. STMAC: [B8:27:EB:72:AC:F8] [60|65 ACKs]
```

Figure 7.9: Perform DoS Attack

Step 6: Validation Finally after performing a DoS attack, we are not able to get a reply from the Raspberry Pi (as shown in Figure 7.10). We can check this by using a command: ping 192.168.43.86, where 192.168.43.86 is an IP address of Raspberry Pi. Thus, we cannot communicate to the home network or automated system using the Internet. In the current scenario, de-authentication of the access point is implemented with the help of transmitting data packets by an

```
root@kali:~# ping  192.168.43.86
PING 192.168.43.86 (192.168.43.86) 56(84) bytes of data.
∏
```

Figure 7.10: Output of Ping Command

unauthorized user. This approach is based on a type of denial of service attack. Several tools and techniques are available to prevent DoS attacks such as identifying the DoS attack in earlystage, over-provision bandwidth, and defending at the network perimeter.

7.6   Research and Challenges

IoT opens the door of opportunities in distinguished application areas such as wearable devices, home appliances, agriculture equipment, medical areas and many more. Although the growth in IoT contributes to distinguished fields, to implement these applications, a large amount of data must be shared on the Internet which is the most vulnerable thing for information security, physical objects of IoT, and other third party information. There are several research challenges in IoT security [400]:

1. Object identification and locating: Unique identification of an object is the very first important issue that can be handled using Object Naming Scheme (ONS) [336]. Locating an object on the Internet can be possible with the help of Named Data Network [259]. Still, there exist several challenges for researchers to provide an efficient approach for identification and locating the objects in IoT.

2. Inadequate Authentication: A traditional authentication mechanism of the user is to provide user name and password, but this is not a sufficient approach to deal with authentication of objects on the Internet.

3. Privacy: User's behavior and activities collected on the Internet may generate privacy issues about the information. There are lots of companies that share that information with third parties for the sake of money. So, it is a great challenge for the researchers to provide privacy for a large information set on the Internet in IoT.

4. Energy Constrained: IoT devices contain limited energy resources due to battery power. An attacker can drain the battery by generating a flood of messages and stop the services for legitimate users. So, this one is another big challenge to deal with small energy IoT devices.

5. Software vulnerability: Software bugs may be the reason of vulnerability in the system. The programmer can be focused on the implementation of the functionalities of the software. It is very difficult to handle security with the mainstream of the software.

6. Access Privilege: After installation of the system, devices cannot request a change in the password or credentials. So, the attacker can access the functionalities of devices in an unauthorized way.

7.7   Conclusion

Internet of Things is one of the emerging areas of this era which helps in connecting things with communication networks and applications. Securing such an outgrowing technology is one of the key concerns. IoT environment is prone to various attacks such as Distributed Denial of Service (DDoS) attacks, spoofing, eavesdropping and malware attacks, etc. In this chapter, a detailed classification of IoT attacks has been proposed. We also proposed a novel IoT security framework in which various attacks, vulnerabilities, and countermeasure techniques have been discussed for each layer. We further implemented a real-time case study on denial of service attack to de-authenticate the access point of home automation application with the help of tools based on the Kali-Linux platform. In the future, we will analyze and investigate different techniques to deal with IoT attacks.

Part III

Engineering and Applications for IoT Cloud Network

Chapter 8

A Novel Framework of Smart Cities Using Internet of Things (IoT): Opportunities and Challenges

Rahul Chauhan

Department of CSE, GEHU, Dehradun, India

Preeti Mishra

Department of CSE, GEU, Dehradun, India

R.C. Joshi

Department of CSE, GEU, Dehradun, India

Cities have shown unprecedented growth in the last few years, bringing some major challenges of adequate services and infrastructure as cities seek to remain sustainable, healthy, employable, clean and safe places for people to live and work. Internet of things (IoT) is a recent communication paradigm in which the devices are equipped with sensors, microcontrollers, transceivers, provided a unique identity (RFID, Barcode, etc.), wireless connectivity and a suitable protocol stack for a seamless connection over a data network. In this paper, a comprehensive 6-layer framework for smart cities is presented in terms of urban information systems incorporating the multiple layers specifying various tasks from sensing to cloud services and providing adequate services to the people. Five major aspects of smart cities are considered, namely, smart healthcare, smart traffic management, smart home, smart grid, noise, and air pollution monitoring. An individual edge server is designed for each mentioned service with suitable sensors and the wireless sensor network (WSN) technology. Each designed WSN sends the sensor data to its edge server on a regular interval of time and further every edge server communicates with each other with a common set of protocols thus creating a smooth information exchange system. Data is being analysed using suitable machine learning algorithm which classifies the data. The future prediction can also be made based on classified data. Besides, various challenges and opportunities have been presented for implementing the proposed framework.

8.1   Introduction

With the increase in population, the cities grow and expand in a similar manner. To meet the ever increasing demand of a human being, a city needs to be smart in terms of infrastructure and living condition. So the demand of a sustainable, clean, safer and smart city is increasing with time, although there is no exact definition of a smart city. However, a smart city includes the optimal utilization of public and private resources and aims to provide the luxury in human life. Nowadays the smart city needs to provide the solutions for traditional problems like traffic congestion, waste management, water and air pollution, smart surrounding and environment and a healthcare sector [388]. A smart city paradigm is shown in Figure 8.1 with all the communication and network technology involved. It can be seen that for a smart city project, cloud computing, edge computing and Internet of things (IoT) play a crucial role. Here the role of IoT is to make everything connected over a network and thus the devices and equipment in the city are supposed to be equipped by the sensors and actuators. The ultimate aim of smart city is to provide the adequate services to society by best utilizing the resources and generate a cost effective solution of the traditional problem in healthcare, environment, etc. The use of IoT is to make a smart city more connected, as IoT is mainly concerned with the connecting of every object and equipment over the Internet with the help of sensor and actuator technology.

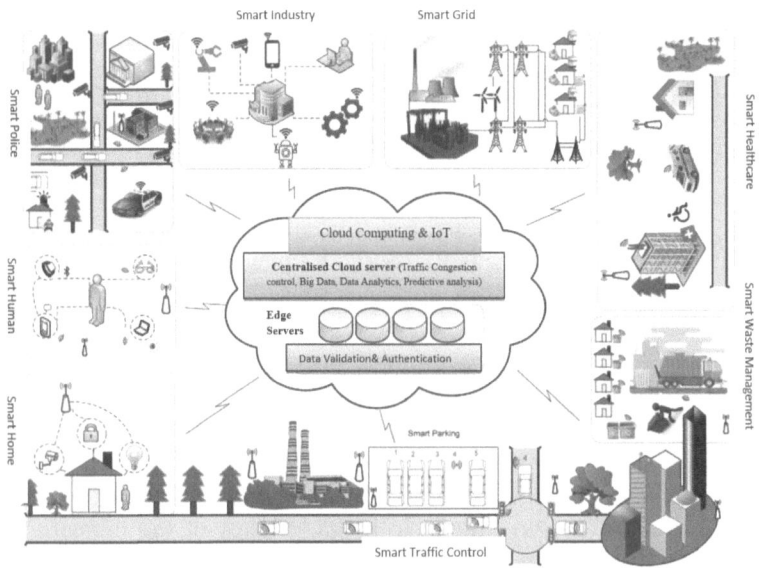

Figure 8.1: IoT information and communication paradigm of multiple services in a smart city

Internet can be utilized in a more immersive and pervasive manner than ever. Thus it creates an opportunity for the stakeholder like the citizens, designer, developer, planner and all to contribute and grow in business perspective also. As the networking grows in the smart city, then is a demand for various communication technology a rises. Table 8.1 shows the common IEEE communication standard for an IoT based application. It is strictly adhering to the standards which are supposed to be followed in IoT based smart city. Whether it is of local network or wide area network, in smart city services, miscellaneous communication standard is required at various level of data flow. In the past a researcher came up with an idea of integrating an IoT with a smart city and so a number of integrating framework a were proposed. Some of the contributions are such as Zanella et.all [388] present in a survey on smart city technologies with respect to a smart city called "Padova." Here they have presented a web service based IoT framework with architecture, protocol and modern connecting technologies. In [181] a survey on fundamental IoT elements with respect to smart city was carried out.

Similarly in [139] the architecture based on top-bottom approach is proposed where the service provider can play a role of the central information unit. It also gave an insight into the smart city communication technologies. But majority

Table 8.1: IEEE communication standard for IoT based applications

	NFC	802.15.4	802.15.3	802.15.1	802.11a	802.11b	802.11g	LoRa
Network Type	Peer to Peer	LR-WPAN	WPAN	WPAN	WLAN	WLAN	WLAN	NB IoT
Network Size	NA	65535	245	7	30	30	30	–
Operating frequency	13.65 MHz	2.4 GHz	2.4 GHz	2.4 GHz	5 GHz	2.4 GHz	2.4 GHz	433, 868 MHz
Range	0 – 0.2m	75-100m	100m	100m	120m	140m	140m	30 miles
Bit Rate	424 Kbps	250 Kbps	55 Mbps	3Mbps	54Mbps	11Mbps	54Mbps	50 Kbps
Modulation Scheme	Modified Miller & Manchester	ASK DSSS QPSK	16-QAM 64-QAM	AFM GFSK 8DPSK	BPSK QPSK 64-QAM	DBPSK CCK DSSS	DBPSK DQPSK OFDM	Spread spectrum
Application Area	RFID	Zigbee 6LoRaPAN	Images & Multimedia	Bluetooth	Wi-Fi	Wi-Fi	Wi-Fi	LoRa WAN

of work is a carried out on network centric framework. So there is a need for data centric integrated framework which can solve the issue of big data and data congestion. Herein the major 6-layer framework is proposed.

8.1.1 IoT infrastructure for smart city

The use of information and communications technologies to make the urban city services and monitoring more aware, interactive, managed and efficient makes it a smart city. Here the basic building block of IoT infrastructure is presented with three different domains, namely, network centric IoT, cloud centric IoT &data centric IoT [388] with correspond to the available information & communication technologies (ICT). Due to the widespread use of wearable technology a new type of domain is created called human centric IoT. As mentioned in the Figure 8.1, smart services require the management & communication technologies to achieve a common goal.

8.1.1.1 Network centric IoT

In terms of service the IoT can be categorised in two different domains. First is 'object' based and second is 'Internet' based. Internet services is the main focus in Internet based architecture while data is being sensed & collected by the objects. In object based architecture [321], data and smart objects become the main backbone and multiple sensing devices are working together to realize the common goal. Networking of sensors and connecting to them on the Internet is required in both cases. RFID, Barcode and WSN are the integrated part of sensing layer which is the innermost layer on an IoT network. RFID is associated with radio frequency identification technology available in the form of active and passive tags. Having a diverse application from security to the unique identification of smart devices in an IoT network. Barcode is associated with the representation of data in the network in machine readable form and this uniqueness is created by varying the width and line space. Just like RFID it is also used for creating the unique identity and has an application in security and authentication. The interconnected network of diverse sensors with the ability to sense and act according to the programmable device is called wireless sensor network (WSN). As the sensor fabrication technology is getting cheaper, more compact and affordable sensors are commercially available for sensing applications.

With the available IPv4 addressing scheme, devices which are already in network are already getting accessed remotely. This is obvious to have a unique identity for the device which is new in the network. So the IPv4 is not sufficient enough to provide unique identification to individual devices which are expected to connect in IoT based smart cities. So the scalability of addressing is not sustainable in the existing network. So IPv6 is the solution of providing unique identity to every device as its addressing range is in the order of 2^{128} bit.

8.1.1.2   Cloud centric IoT

The 'Internet' centric and 'things' centric are two major attributes of an IoT based smart city framework. Cloud centric IoT is mainly focussed on cloud computing paradigm wherein the cloud holds the centre stage. Ubiquitous computing and sensing can play a major role in the smart city framework. So the combined framework of ubiquitous sensing and cloud computing is more viable in smart city perspective [182, 149]. In cloud centric IoT the sensing layer offered its services to the cloud server thus reducing the burden of data congestion and data analytics. Thus the majority of data filtering and data classification and data analytics are carried out at cloud server level. At cloud server level the miscellaneous tools of analytics, knowledge base of an artificial intelligence expert and machine learning tools can combined to create a more meaningful information and knowledge. In the smart city application, both public and private clouds are required with authorization in multiple applications. And this integration can be possible by Aneka which is a .NET based application development platform. There are various other platforms also for development like Microsoft Azure, Think speak by math-works and IBM Watson. They can provide the service of data analytics. A single seamless framework is, however, more essential in order to get sensed information. Such single seamless framework is proposed in figure 8.4 with sensing layer to the cloud computing layer integrating various smart city services together. In figure 8.2, a cloud centric IoT platform is shown with cloud at centre stage. It has been estimated and observed that the IoT sensors also play a vital role in generating a large amount of data. Scalability of sensors is dependent on the application and also the commercialization of sensing technology. To cover the large geographical area, a wireless sensor network is also going to play an important role wherein the data to the central cloud is being transferred by the help of the gateway. As shown in the Figure 8.2, it is a bidirectional control and information exchange process with the client application as the subscriber service [134, 214].

8.1.1.3   Data Centric IoT

By the year 2020, 600 zettabytes of data per year will be generated, as estimated by CISCO. Main source of this data will be the Automobile industry

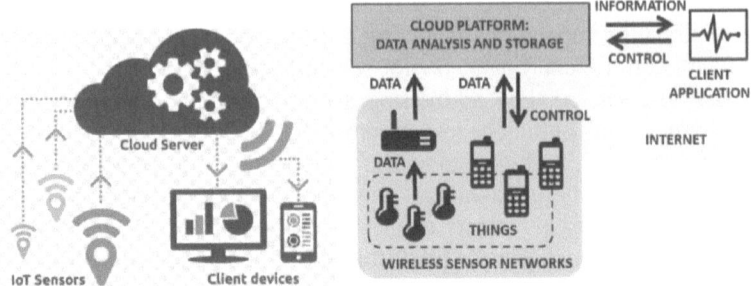

Figure 8.2: (a) An example of Cloud service for IoT, (b) A Generalised cloud centric IoT platform

and energy and manufacturing sector. The radical transformation in the automobile industry forced us to accept the industrial revolution 4 and thus create a widespread sensor network and sensors data. Data centric IoT mainly emphasis on the data flows starting from collection to the processing and visualization. IoT sensors collect the data and this raw data will be processed for meaningful information. For smart city application the adaptability and robustness in the data analytics algorithm is highly required as there is a large variation in data generated in IoT based smart cities. The data can be analysed at three different machine learning stages for the central cloud server as well as the fog level (edge level).

8.1.1.4 Human Centric IoT

The commercialization of smart sensors and wearable technology creates an opportunity for the scalability of human centric IoT infrastructure. Starting from the measurement of heart rate, calories, number of steps taken, energy, speed, motion and many more the wearable technology is everywhere. Easily accessible fitness band are generating an ample amount of data to get knowledge about human health in smart cities. Herein the smart devices, gazettes, and smart phone also play an important role in communicating the information captured by the person's smart wearable [130]. Human-centric IoT applications are relevant to a wide variety of needs and innovative technologies. The contributions of this chapter are as follows:

- To develop a comprehensive integrated 6-layer data oriented framework for smart city using IoT.

- To integrate all communication technologies associated with smart city.

- Provide an insight into the challenges and opportunities in realising a smart city.

- To provide a case study in terms of prototype development of smart parking system and IoT centric pill bottle for better understanding of practical situations.

The rest of the chapter is organised in the following sections. The section 9.1 summarises the IoT infrastructure based on three different approaches, namely, network centric, data centric & cloud centric. Next section describes the smart city hierarchy wherein the network requirements, communication protocols and features are mentioned. In section 8.3, a comprehensive 6-layer smart city framework using IoT is proposed starting from data acquisition layer to the application layer incorporating various layers of data abstraction. In section 8.4, a case study is presented in terms of smart parking system and an IoT based pill bottle to validate the practical scenario. In the last section various opportunities and challenges are discussed that are associated in realizing the smart city.

8.2   Smart City Hierarchy

Smart city concept involves the working of multiple communication technologies, protocols, service providers, heterogeneous networks and efforts to achieve a common objective. In this section hierarchy of a smart city is presents that summarized the miscellaneous communication protocols, service providers, variety of networks, requirements and services offered.

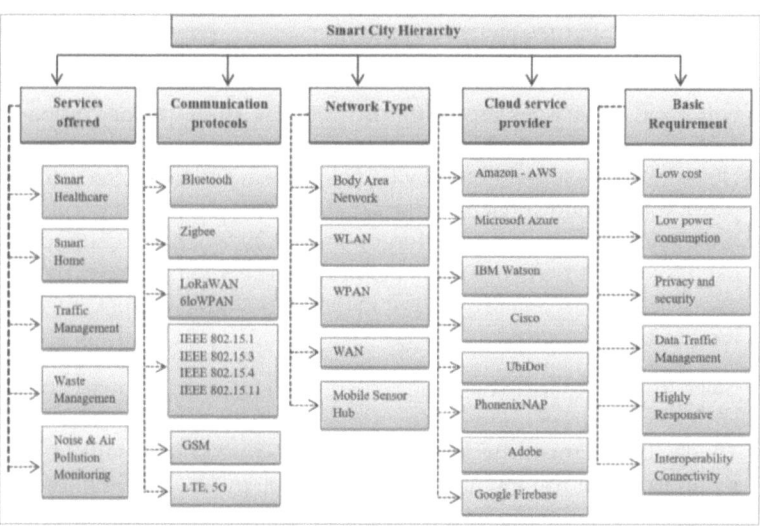

Figure 8.3: Smart City Hierarchy

(a) **Offered Services:** Smart city facilitates the smart services to the people involved with smart healthcare, smart home, smart traffic management system, waste management, noise and air pollution monitoring. All the smart wearable's sensors can contribute to the smart health monitoring. Starting from the measurement body parameters by smart sensors to value analysis, future prediction of disease. Similarly, the use of smart sensors can result in smart homes wherein all the appliances can be remotely accessed [169]. The use of a density based traffic light control system and smart parking can solve the problem of traffic and it can be an aid to the smart city. Smart dustbin with cloud connectivity can easily solve the problem of waste management [169]. The use of GPS and GPRS can reduce the burden of regular monitoring and ovelead dustbin. And the multiple noise sensor and harmful gas sensor can be installed at crowded locations to cover the maximum geographical area and monitored over a time for alert generation.

(b) **Communication protocols:** Bluetooth, Zigbee, 6LoWPAN, LoRaWAN and IEEE 802.15 are some prominent communication technologies associated with the IoT smart city framework. LTE can be utilized in healthcare. The interoperability between several operators is being supported by the standardization of LoRaWAN protocol.

(c) **Network Type:** There are multiple network topologies used in a fully accomplished and autonomous smart city network. Short range communication like local area network (LAN) and personal area network (PAN) depend upon the application in smart city. Wide area network (WAN) and Metropolitan Area Networks (MANs) are the popular choice for the healthcare and waste management application.

(d) **Cloud Services:** With the growth in IoT research with respect to industries, better cloud services are being provided by some major companies like AWS by Amazon, Watson by IBM, Thingspeak by mathworks, CISCO, UbiDot, Adobe and firebase by Google. These entire service providers can provide the cloud services for smart cities with the subscription. And even it depends upon the application of smart city where it is supposed to be used. Like for emergency services, where the response time is more critical as compared to the data analyzation then one should prefer Google firebase instead of Thingspeak.

(e) **Basic Requirement:** The basic requirement from every smart city services is that it must be a low power consuming device that should be used. Trying to minimize the latency and data congestion should be avoided by appropriate routing algorithms. Technology should be affordable and privacy of data should be conserved by using appropriate data encryption algorithms. As the proposed framework is an integration of heterogeneous smart city services, so the system should be interoperable.

Table 8.2: IEEE communication standard for IoT based applications

Associated Technology	Operating Frequency	Speed Data Rate	Power requirement	Range	Protocol	Approximate delay	Specific Service
Bluetooth	2.4 GHz	25 Mbps	0.5mW-100 mW	10m-100m	IEEE 802.15.1	100ms	Smart Home
Zigbee	2.4 GHz 915 Hz	250 kbps	2.7 V – 3.6 V 45mA – 120 mA	100m	802.15.4	16-18ms	Smart Home Smart meter Smart Healthcare
Wi-Fi	2.4 Ghz	54 Mbps 6.7 Gbps	20mW-4W	100-150m	802.11n	45ms	Waste management Noise monitoring
6LoWPAN	2.4 GHz	250 Kbps	3.3 V	100m	802.15.4	15ms	Smart Meter
LoRaWAN	433,868, 915MHz	50Kbps	Low	2-5 km	NB-IoT	–	Waste Management
GPRS	850,1800, 1900MHz	80-384 Kbps	4.5V – 12V	5-30km	SNDCP	1.5 s	m-health, smart meter
3G	850 MHz	3 Mbps	High	5-30 km	BICC,	100ms	m-health, Traffic management
WiMax	10-66 GHz	30-40 Mbps, 1Gbps	300W	50 km	IEEE 802.16m	50ms	Smart Grid, Smart Meter
LTE	750,1900, 2500MHz	1Gbps	High	5-30 km	–	5ms 5ms	Mobile health, Traffic Management
5G	6Ghz-24GHz	1Gbps – 70 Gbps	High	< 30 km	–	< 1ms	Traffic Management, Smart Grid

8.2.1 Associated communication technology for realizing smart cities

IoT offers the diverse application with respect to smart cities. Interoperability, privacy, security, flexibility, large coverage and range, and low cost and low power consumption are some inherent expectations from IoT based systems.

The safe and secure IoT communication is significantly important to ensure the data integrity and security. Table 8.2 summarises the associated technology with IoT based smart cities and its multiple parameters starting from power requirement to the range, protocols, data rate and specific service associated. The short range communication technology like Bluetooth and Zigbee can be used for sending the data to the local server/edge server through gateway. For instance, smart traffic management required a modelling such that the traffic congestion should be avoided. Here the network latency is not acceptable and the system needs to be highly responsive. For such application the data can be sent to the local server using Bluetooth or Zigbee. Further with the help of Wi-Fi the data can be sent to the central cloud server. This integration of communication technology can improve the performance of the system. For the application where the bandwidth and speed requirement is high, like the CCTV cameras for traffic monitoring, 5G is preferred for high data rate.

8.3 Proposed Framework

Developing a common architecture for IoT in context to smart cities is a challenging task, mainly because of heterogeneous services and variety of devices and services involved in the smart cities. In figure 8.4, the 6-layer framework is proposed with sensing layer, data abstraction layer, base station layer, edge server layer, cloud server layer and application layer. Here 5 different attributes of smart cities are considered, namely, smart healthcare, smart home, smart traffic management, smart grid and noise monitoring. Each attribute is classified in the six-layer architecture starting from sensing layer to the application layer.

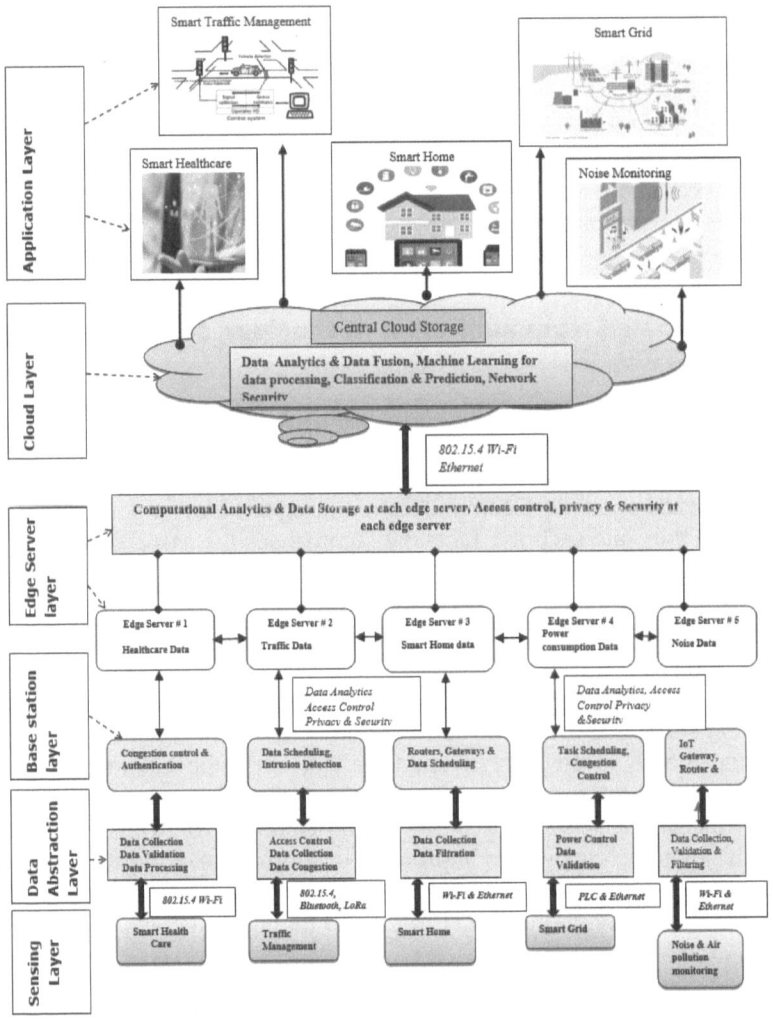

Figure 8.4: Proposed novel framework for smart cities incorporating various layers

Table 8.3: Possible types of sensors used in Smart city application using IoT

Type of sensor	Features	Applications
Temperature sensor (LM-35)	Measurement of Home temperature	Smart Home
LDR	Use in Smart Traffic light and Measure amount of light in home	Smart Home, Traffic management
Motion Sensor (PIR)	Detect motion in home	Smart home, traffic management
Humidity sensor (DHT -11)	Measure humidity level in home	Smart Home
ECG Sensor (AD8232)	Measure ECG signal through wrist	Smart Healthcare
Body temperature sensor	Based on surface contact, measure body temperature	Smart Healthcare
Thermistors	Used in smart grid and smart home to measure smart environment temperature	Smart Home, Smart Grid
Infrared sensor	Used in smart parking and density based traffic light control system	Traffic Management
Smoke sensor (MQ-2)	Detect fire caused in home	Smart Grid, Smart Home
CO2 detection sensor (MH-Z19)	Check the content of harmful CO2 emission with respect to oxygen level	Air Pollution monitoring
Smart Thermostat	Measure temperature in Smart Grid	Smart Meter, Smart Grid
Noise/Loudness sensor (LM2904)	Measure the noise intensity in smart city	Noise monitoring
Transducer	Used to convert the power line signal voltage into electrical signal	Smart Grid Smart Grid

8.3.1 Sensing Layer

Sensing layer is mainly concerned with the data acquisition for the framework, as every layer in the framework depends upon the data to work. So the data acquisition is a crucial step and, based on the services of a smart city, the sensor requirement is also variable. Table 8.3 summarised the possible sensors used in smart city services. Sensing layer ensures that the correct and clean information to the above layers and there should be a smooth data flow in the network.

These variations create an issue of heterogeneous data integration and data fusion. For instance, smart home mainly requires temperature sensor (Lm-35) for home temperature measurement, LDR for automatic room light control, motion sensor for any suspicious activity monitoring & smoke sensor for any fire situation.

8.3.2 Data Abstraction layer

The data generated by sensing layer is transferred to data abstraction layer and it is responsible for the data collection, data validation and data processing. This layer is also responsible for the access control, power control and data filtering. Data is sensed by the sensing layer and data variance is quite high and it is generated by multiple attributes of smart city. Not every data generated by the sensing device is meaningful and so the majority of data is discarded or filtered out because it is not considered enough for retention. This layer is required to avoid the burden of data congestion on edge and cloud server. And only meaningful data is being transferred to the edge server. Some operations/processes which are supposed to be taken at this layer are mentioned below.

(a) **Data Collection:** The regular and scheduled data collection is required in every application of smart city. Like in healthcare, the regular data of body parameters should be collected and it should be scheduled in such a way that it doesn't create the problem of big data. In a smart city application data collection is an important stage because all the above layers directly depend upon the data send by the abstraction layer.

(b) **Data Validation:** Clean and clear data is highly required in every application of smart city. Specifically the data related to the body parameters should be clean and clear; otherwise it can create a problem in medical diagnosis. It should be precise and accurate in every aspect. As the sensor node generates the heterogeneous data, validation is considered to be the important step. Data validation is often performed prior to the ETL (Extraction Translation Load) process. A data validation test is performed so that analysts can get insight into the scope or nature of data conflicts.

(c) **Data processing:** As mentioned earlier not every collected data is meaningful and useful. So some initial pre-processing and filtering is carried out at this stage to get clean and meaningful data. A sequence of operation can be performed to generate a data which can be analysed. For instance, a regular body temperature monitoring of a patient in smart healthcare is required and only the maximum and minimum body temperature is needed. Then at this stage maximum and minimum value should be taken and the rest can be discarded to avoid the unnecessary data congestion.

(d) **Access control:** In the physical system access control is used to provide the security and authentication to the system. In every application, whether traffic management, smart grid or healthcare, authentication is required to avoid misuse of private data. Technology like Radio frequency identification and detection (RFID) Biometric identification can be used to provide the authorised access in smart home and smart grid.

(e) **Power Control:** Due to the heterogeneous data generation in the smart city, power requirement is also variable in every application. Some applications work on low power and some require high power to operate. As in smart home and smart healthcare low power operating sensors are used whereas in applications like smart grid, high power consuming sensors are used. So the power distribution should be a challenging task in smart city.

8.3.3   Base station layer

This layer is mainly responsible for the task related to the data scheduling, congestion control in a network, intrusion detection, routers and gateway in a framework. This layer is responsible for sending the filtered data to the edge server and ensures the data scheduling and smooth transition of data flow.

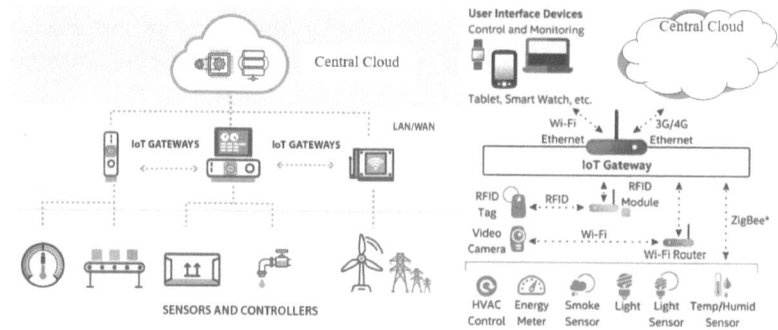

Figure 8.5: The Base station layer integrating the sensors, actuators with the Gateway and edge server

Figure 8.5 demonstrates how the sensors, actuators and controller can be used as an integral part of a router and gateway network. The selection of energy efficient routing algorithm is also critical to best optimise the power with more integrity in the data flow.

(a) **Data Scheduling:** For successful exploitation of a broadcast medium and data transmission the efficient scheduling algorithm is highly required. While accessing the data through Road Side Unit (RSU), data scheduling becomes an important issue. Therefore, to deliver the messages to the recipient properly and accurately, scheduling algorithms have to be emphasized [46]. There are various scheduling algorithms exist like request based scheduling algorithm, time based scheduling algorithm, deadline based scheduling algorithm and hybrid based scheduling algorithm. Here hybrid based algorithm can be used because it can deal with parameters like size, deadline of request, and time which is obvious in this integrated framework [334].

(b) **Congestion control:** In network there is always a possibility of data overflow. As in smart city the amount of data generated is huge and random. So it should be considered that meaningful data should not get overflow. Leaky bucket algorithm and token bucket algorithm [351] can be best utilized in smart city application to avoid network congestion. Even for smart city application token bucket algorithm is preferred over leaky bucket because it is more flexible.

(c) **Intrusion Detection:** For smart healthcare and smart home application security and authorization is a critical issue. Any intrusion in the network should be monitored and corrective measures should be taken. An intrusion prevention system (IPS) also monitors network packets for potentially damaging network traffic. But where an intrusion detection system responds to potentially malicious traffic by logging the traffic and issuing warning notifications, intrusion prevention systems respond to such traffic by rejecting the potentially malicious packets. Network intrusion detection system (NIDS) should be strategically used to avoid any unauthorised access.

(d) **Routers and Gateways:** To cover large geographical area signal repeaters are required in smart grid, noise and pollution monitoring system and traffic management. Routers can solve this purpose. Besides that collected information from sensors, sensor nodes need to be sent on the cloud. So gateway is required to send it to the cloud server as it can provide the data routing as well as the Internet connection. The data integrity needs to be preserved while communicating in a smart city. So to meet the demand of smart city an adaptive routing algorithm is required. Smart and self-organising routing algorithm (SSRA) selects the best route for data packets [153]. Energy efficient consumption can be observed in SSRA.

8.3.4　Edge server layer

To reduce the burden of central cloud server and to overcome the problem of big data, every smart city attribute has its own edge server. They are just like the central cloud server but with limited computational power and access and lie close to the edge devices. The main objective of using edge server is to avoid the latency in the emergency situation like fire alert and medical alert. Smart city put a unique challenge to the database as it increasingly produces a large amount of data. So some processing and analytics need to be done at the network edge, closely with sensors and actuators used in the smart city application. In figure 8.6 the edge computing platform of smart city application shows that all the services of smart city are connected to a dedicated edge server and all the edge servers are interconnected for smooth transition of information. As mentioned earlier the load of main cloud server is being distributed in multiple edge servers [319]. For instance, the data generated in healthcare application first need to transfer on the edge server and some data analytics need to be performed with the tools available on edge server. In case of emergency, data can be easily accessed remotely with minimum latency. Also the data which doesn't require urgent attention can be further sent to central cloud for predictive analysis. Like for a diabetic patient, blood glucose level needs to be monitored for a large number of days to get a prediction of disease.

For such case the analytics should be carried out at the central cloud. The distribution method and data transfer method are optimized according to the

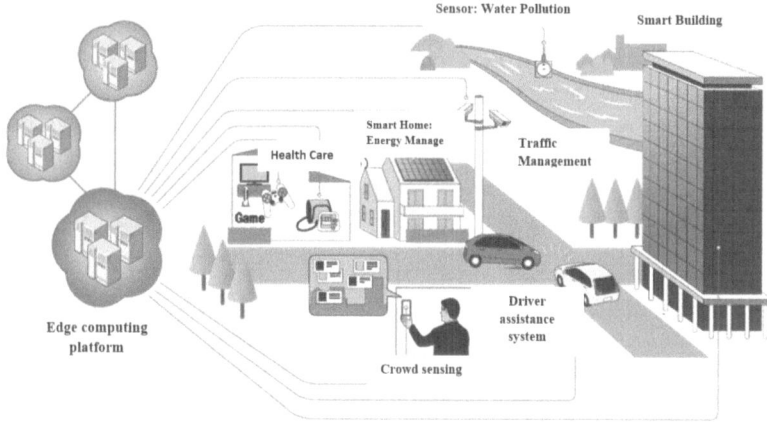

Figure 8.6: Smart city application in edge computing platform

user's environment, such as device capability and the networking environment, such as fixed or mobile. For example, the processing of objects, such as images and sounds, is dynamically allocated to devices or edge servers to suit the user's environment, and resolution and frame rate are controlled to suit the networking environment. Edge server also needs to ensure the privacy and security issues of the network.

8.3.5   Cloud computing layer

Although every attribute of smart city is equipped with dedicated edge server for the cloud storage and data analytics, still it has limited computational power and less data space for the application which required a large amount of sample data to classify and predict. For instance, with the prediction of any heart disease of a patient based on the collected data one needs to generate a database of huge amounts of dataset. So that database can generat in central cloud and by machine learning algorithm. *Data fusion, data classification, prediction & network security* are some functions and issues concerned with the cloud computing layer. There are multiple data sources in smart city paradigm, so at cloud server, heterogeneous data accumulated and thus data fusion technique need to be applied to integrate multiple data to generate more consistent, regular, accurate and meaningful data [118]. Similarly the data from multiple sources need to be classified for more meaningful information. Here the adaptive machine learning approach should be used for various levels of data classification.

8.3.6 Application layer

In the proposed framework 5 different attributes of smart cities are considered, namely, smart healthcare, smart home, smart grid, smart traffic management, and noise & air pollution monitoring. They are many more attributes that can also be possible but the feasibility and accessibility need to be considered. Mainly this layer focussed on miscellaneous smart city services and focussed on the infrastructure by computing the data available at cloud server.

(a) **Smart Home:** The use of smart sensors and actuators make it possible to design the smart connected home with the *information communication technology (ICT)* paradigm. Temperature controlled environment, energy efficient building, and Internet connected home appliances can contribute in designing the smart home.

(b) **Smart Healthcare:** The use of an Internet connected Pill bottle, wherein the daily monitoring of medicine intake can be recorded in an online database, and use of sensors for body parameters measurement can alert the physician in an emergency situation, which makes a smart healthcare system in smart cities.

(c) **Smart Grid:** In smart grid the traditional electric grid is integrated with the modern digital communication system and sensors to best utilize the power consumption of a city. The use of smart meters where the power can be shared with the local grid and some other user makes it a more adaptive system in modern days. Ultimately every service in smart city needs to depend upon electricity to operate; so its optimal utilization is highly required.

(d) **Traffic management:** Smart traffic light control system and smart parking system are two major aspects in smart traffic management. Use of CCTV cameras and traffic density sensor and crowd sensing can help in identifying the potential traffic in an area at a remote location.

(e) **Noise and Air pollution monitoring:** For the measurement of noise pollution, noise sensors can be used at some identified location and, similarly, for checking the concentration of harmful gases in the air, gas sensors are used in a wireless sensor network to cover the large geographical area of the city.

8.4 Case Study

For the validation of smart city infrastructure, here the case study is presented for smart traffic management, using a smart traffic control pedagogy and smart parking approach, and for smart healthcare in terms of an Internet connected pill bottle.

8.4.1 Smart traffic management

Real time public traffic analysis improves the quality of life and increases the efficiency of transportation and logistics without human involvement through IoT. It increases the comfort level and saves the resources. The figure 8.7 shows how a smart parking approach works with the integration of Raspberry Pi cameras, density based sensors and IoT. It can be proved as a robust smart parking approach and can solve the problem of vehicle chaos in an efficient manner.

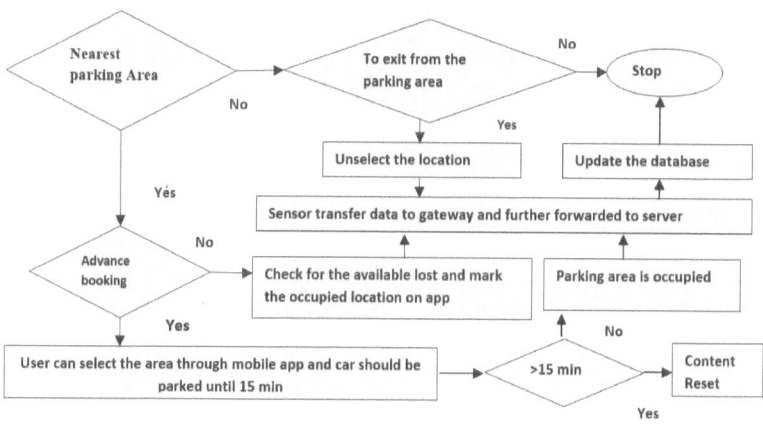

Figure 8.7: Smart parking approach using sensors and IoT

8.4.2 Smart Healthcare

In smart healthcare multiple body parameters can be measured for better diagnosis of disease as well as for better prediction of a disease. Here a concept of a smart pill bottle is presented. As people enter into old age, memory starts degrading and often people forget to take the prescribed regular medicines. Imagine yourself receiving a phone call or a text message from your pill bottle, which tells you that you have just missed the pills you are supposed to take or you have taken the non-prescribed dosage. A smart pill bottle can do all these tasks by integrating the Internet and cloud services with the sensor equipped pill

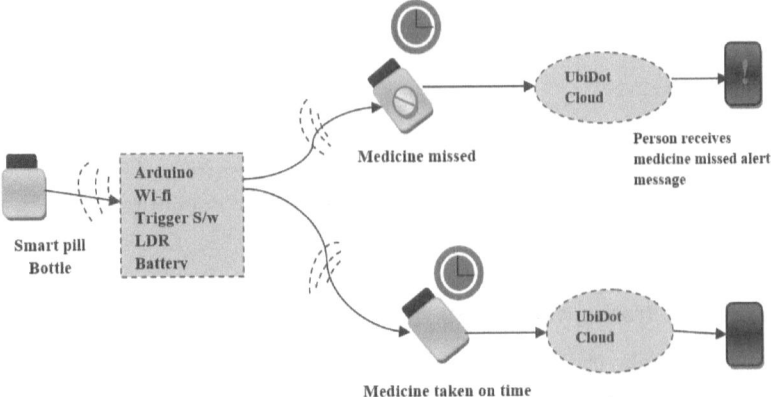

Figure 8.8: Systematic flow of data in an IoT centric smart Pill Bottle

bottle. Figure 8.8 represents the working model of a pill bottle with all necessary hardware and cloud requirements. Here the *UbiDot cloud* service is used.

LDR sensor is sensitive to light. When it receives the light as signal, it responds to the Wi-Fi module connected with the Arduino development board. LDR sensors attached to it determine whether the person is taking pills in time or not based on the light condition whenever the bottle cap is opened. When the pill count goes below the threshold value (minimum count of pill) a message of purchase new pills is delivered to the pharmacist through the Ubi-Dot cloud. In case person forgot to take medicine in prescribed time then an alert is generated and a message is sent to the concerned person. So this technology of pill bottle actually can avoid unnecessary diseases which may occur due to inappropriate and irregular dosage of medicines.

8.5   Open Challenges and opportunities

(a) **Business model:** To improve business, business leaders aim to provide the solution of big data. Formation of a sustainable and integrated business model for smart city is a challenging task. For a smart city planner it is a big challenge to have an estimate of an amount of data generated in a smart city [142]. Expansion of business in terms of IoT devices and sensors also creates an opportunity for sensor design engineers. But scalability remains a concerning issue.

(b) **Sustainability:** In various smart city applications, there will be a real time interaction between the devices and communication technology. Large amount of data inflows and outflows into the system thus creating an issue for sustainable development. As more and more dependency will be on the edge devices to generate a clean and clear data for the system. The integration of big data and IoT can be best utilized in a sustainable smart city [207].

(c) **Connectivity issues in IoT:** There are billions of devices connected together in a smart city and they can be expanded in a regular manner to meet the city need. Connectivity in each and every available device in a system is a challenging task. At various levels there are number of communication technology like Bluetooth, Zigbee, LoRaWAN, Wi-Max, Wi-Fi & Ethernet. So to ensure a smooth and safe connectivity among all these technologies is again a big challenge. Network breach and network failure can accidentally create a system failure at any time. So to provide the smooth uninterrupted network required is a high priority in smart cities.

(d) **Privacy, security and trust:** There are ample amounts of sensors and devices connected in smart city and providing security at various levels of data abstraction and in network is again a challenging task. According to the researcher [71] 70% of devices are prone to attacks that are being used in the IoT because of poor authorization scheme and weak encryption in data communication. Attacks which are based on vulnerability and threat need a special assessment. A proper risk assessment should be done. ENISA is such a framework [215] proposed by some researcher to identify the attack.

(e) **Lower power consumption:** Life of sensing devices and processing elements need to be considered while using them in smart city applications. Mostly the devices used in IoT are small sensors and actuators which operate on a battery. So the power should be optimized in such a way that the battery lasts for a longer duration. Power consumption can also be controlled by the adaptive routing algorithm which can put the sensor node and WSN in sleep-mode when not required. In this way the device life can also be used and can operate on single battery for a longer duration. Ultimately reducing the overall cost of system.

(f) **Big data & Data Integration:** A massive amount of information and data is generated by devices used in smart city. This data includes the data from healthcare sector, smart home, noise and traffic sensor data, and smart grid data. Thus there is a large variety in data and thus its integration throughout is a challenge to the designer. To have clean and clear data with proper format is a big issue. Besides that, it creates the problem of big data. But it also creates an opportunity for the big data analytics enthusiast to generate a meaningful information system for a smooth data flow in smart city.

(g) **Cost of technological acceptance:** The big question is are we ready to pay for all these smart services in a smart city? How much cost is acceptable to have these entire enhancements in traditional city services? It requires a huge infrastructure to even implement for a single smart city services, for instance smart home requires installing sensors and smart meters in each and every house in a city. The challenge is who will pay for these, the customer or the city administration? All these issues need to be considered before applying smart technology in the smart city.

8.6   Conclusion

In this chapter, a 6-layer framework for smart city is proposed, starting from sensing layer to the application layer. An insight to the smart city services is discussed in the paper with all possible communication technologies that are supposed to be used in smart cities. Due to the variety of services involved in smart city, an integrated framework creation for a city is a challenging task. There are a large variety of sensors and devices used in the smart city application which can create the problem of data fusion and integration to the data engineer and designer. A case study presented on smart traffic management and smart parking can solve the problem of traffic congestion. The prototype of smart pill bottle has been designed for the single user and can be extended for multiple users per demand. Although the smart city framework seems to be effective as far as the technological and business point of view is considered, it still puts on a lot of challenge to design engineer. There are a number of issues which require major attention like authentication, security, privacy, big data, data integration and many more. It also creates an opportunity for those skilled in data analytics, hardware design and network design.

Chapter 9

Interoperability and Information-Sharing Paradigm for IoT-Enabled Healthcare

Brian Desnoyers[†]

MIT Lincoln Laboratory

Kendall Weistroffer[†]

MIT Lincoln Laboratory

Jenna Hallapy[†]

MIT Lincoln Laboratory

Sandeep Pisharody[†]

MIT Lincoln Laboratory

As the line between clinical and personal health devices is blurred with new personal health technologies, there is a need for secure and reliable integration between enterprise Internet of Things (IoT) networks, private cloud networks, and personal connected health devices. This chapter lays out foundational IoT and cloud information-sharing requirements for healthcare, reviews existing and potential approaches to facilitate this integration, and analyzes methodologies for achieving heterogeneous data interoperability between various IoT sensor ecosystems. We describe the information-sharing requirements for a healthcare system infrastructure, and the corresponding security effects on infrastructure-as-a-service (IaaS) and private cloud solutions for data management. The integration of data from personally used IoT sensors, such as smartwatches and fitness trackers, with clinically collected information accessed by medical professionals introduces further security challenges and ethical issues regarding data ownership, efficient data sharing, and privacy. Many of these challenges emerge from traditional medical record access patterns, such as allowing delegation of data access controls during emergency care and the sheer number of personnel accessing medical data for consultations and support, potentially without the full awareness of the patient. The information sharing requirements for a modern healthcare infrastructure, based on IoT endpoints for data collection and cloud computation and storage, include efficient data sharing, access auditing, data filtering and transformation, as well as customizable delegation of data access management responsibilities. We then enumerate various information-sharing approaches to meet these unique demands for IoT and cloud integration in the healthcare field, along with the associated efficiency, availability, security, and ethical consequences of each approach. IoT devices add to data sharing challenges that exist today, starting with inconsistent connectivity interfaces, such as WiFi, Bluetooth, and emerging communication interface technologies, like 5G. The efficiency and adequacy of these approaches will be examined in further detail through the lens of scenarios and dilemmas that may be common in future integrated IoT and cloud healthcare environments. This review includes both existing and potential implementation approaches for IoT and cloud data sharing, providing specific examples of proposed and established systems with their benefits and limitations. Finally, we analyze approaches for achieving interoperability

---

† Work submitted in this manuscript was done by the authors as independent researchers. The opinions, interpretations, conclusions and recommendations are those of the authors and are not necessarily endorsed by MIT Lincoln Laboratory.

between various ecosystems of IoT sensors to facilitate heterogeneous sharing of relevant IoT-derived health data between patients, healthcare providers, payers, and other authorized parties. The sharing of big data collected from IoT sensors in real-time will require specialized approaches to achieving this data ubiquity beyond requirements necessary for traditional medical records. With existing interoperability challenges between electronic health record (EHR) platforms, IoT sensor data must integrate with public cloud environments to enable improved clinical decision-making and oversight. This analysis will specifically discuss the data ubiquity, availability, and performance consequences for each interoperability approach while enumerating general best practices for integrating, aggregating, and sharing heterogeneous data across multiple IoT ecosystems and cloud environments.

9.1   Introduction

With the advent and proliferation of Internet of Things (IoT)-based health devices around the world, the distinction between clinical and personal devices is becoming increasingly blurred, resulting in unique information-sharing challenges. The global infrastructure of IoT-based health devices consists of a large number of connected legacy medical sensors, IoT-based personal health devices, and software applications that generate vast amounts of medical data that need to be processed, correlated, and analyzed in near real-time. Given the extensive amounts of data, collecting and aggregating the appropriate data from these systems and performing required data processing and computation requires secure and reliable integration of enterprise IoT networks, public and private cloud networks, and personal-connected health devices.

Healthcare is a relevant case study of IoT-cloud network management because it poses several relevant challenges:

1 the existing legacy of medical records and electronic medical records

2 the blurring distinction between medical devices and consumer devices used in healthcare

3 the balance between privacy and decentralized immediate access to data across healthcare providers.

In this chapter, we detail requirements for secure and reliable data management and the integration of IoT-enabled healthcare.

The chapter is presented as follows. Section 9.2 and Section 9.3 discuss two related efforts that inspire requirements for improved interoperability and contribute to the changing paradigm within healthcare: mobile health (mHealth) and precision medicine. Section 9.4 presents and discusses the challenges to address regarding the ownership of healthcare IoT data, and other integrated medical data sources. Section 9.5 presents the challenges and strategies necessary for data sharing in an IoT-enabled healthcare ecosystem. Finally, Section 9.6 provides a general architecture for interoperability, and then discusses the current state of health record interoperability standards.

9.2 Mobile Health and the Internet of Medical Things

In today's world, mobile technology is ubiquitous. Handheld devices such as smartphones and tablets provide access to information and communications across the world. According to the Pew Research Center, an estimated 94% of adults living in advanced economies own a mobile phone, with the numbers expected to increase within the next few years [357]. This growth in both users and the use of mobile and wireless technologies over the last few years promises a rise in new opportunities for the integration of mobile health technologies. Mobile health, or mHealth, provides users with mobile self-care through the use of consumer apps, devices, and connections that enable users to capture their own health data [246] and receive personal health interventions. Currently, mHealth provides a broad range of services to users, including survey and questionnaire delivery [359], real-time habit recognition and adherence support [63, 138, 276, 374], and pervasive sensor data collection [299, 376]. Although there is no standardized definition of mHealth, we have adopted the definition proposed by the World Health Organization in this chapter:

> mHealth is the use of mobile and wireless technologies to support the achievement of health objectives [257].

In addition to mobile technology, the emergence of affordable, wearable devices has continued to create new opportunities for mHealth. These wearable devices (commonly referred to as "wearables") provide users with a convenient means to monitor and manage personal health and connect to healthcare providers via telehealth (e.g., remote patient monitoring) [263]. Although a vast majority of general-purpose wearables lack specialized health sensors, they have technology components that can provide functionality akin to that of health sensors, such as motion measurement, body tracking, body balance assessment, and pattern recognition [152]. In a broad sense, this ecosystem of connected IoT-based health devices has been termed the "Internet of Medical Things" (IoMT).

Research studies regarding the efficacy of mHealth interventions and outcomes are limited with current evidence showing mixed results [251]. Thus, continued initiatives to conduct systematic studies on the effectiveness of mHealth are essential in determining whether health-related IoT devices are engaging and providing actual value to users, rather than simply collecting data. If mHealth devices are failing to provide tangible value outside of data collection, adherence may not be sustainable for the general population [99, 263]. Only after overcoming this challenge in future mHealth deployments can these devices provide users with low-cost and real-time mechanisms for the assessment of personal, clinical, and public health through the collection and analysis of movement, imaging, behavior, social, environmental, and physiological data [99].

9.3   Enabling Precision & Personalized Medicine

Evidence-based medicine is the practice of integrating the experiences and knowledge of a healthcare provider with external clinical evidence and patient needs [317, 318]. This integrated evidence comes in many forms with the responsibility for seeking out the best external evidence falling, at least partially, on healthcare providers. The "gold standard" source of external evidence is the randomized controlled trial (RCT) [318, 349], that often evaluates treatment effectiveness on the population scale through clinical epidemiology. Integration of external data is practiced by many healthcare providers today and has shortened the gap for new clinical research to be widely utilized in medical practice [252]. However, a significant challenge remains as healthcare providers must still decide how new studies pertain to their individual patients at the time of care [252]. Eric Topol uses the example of widely prescribed statin drugs for preventing endpoints, such as stroke and heart attack, to illustrate this challenge [4]:

> Instead of identifying the 1 person or 2 people out of every 100 who would benefit, the whole population with the criteria that were tested is deemed treatable with sufficient, incontrovertible statistical proof.

At the time of Topol's writing, common evidence-based practice often involved prescribing statins for large portions of the population, such as elevated calculated LDL cholesterol levels [4]. In the future, this approach could expand with the widespread use of polypills, such as those containing a statin along with aspirin and folic acid [234]. While this approach could be considered better than the alternative of a non-evidence-based practice, providing care based on population risk factors determined from a limited set of data can expose a population to unnecessary side effects (the use of statins has been associated with diabetes mellitus, liver damage, muscle damage, and central nervous system complaints [358]) while adding financial burden to the healthcare system.

A simplified practice of personalized medicine has been commonly deployed through pharmacogenetics to tailor drug prescriptions based on genetic markers [211]. However, it is the integration of IoT and mobile data sources that will allow for the inclusion of behavioral (e.g., activity patterns, habit detection) and environmental (e.g., noise exposure, air quality) data into this process. For example,

Joshi et al. [184] describe the integration of IoMT data sources into the neonatal sepsis prediction process.

Current clinical practice has utilized coarse population models to improve patient care. One such example is antibiograms, which identify local patterns of antibiotic resistance. Clinicians currently apply these localized resistance profiles to best identify the antibiotics to prescribe to patients [208], such as within a hospital setting. A precision medicine approach that improves upon this practice might incorporate additional features, such as social network and location check-ins, and provide these insights for the individual patient. Another example of such a clinical model is the Breast Cancer Risk Assessment Tool (BCRAT), which is a model used clinically to determine breast cancer risk based on factors such as age, race, and family history [268]. This BCRAT risk score can be used to determine recommendations, such as secondary prevention screenings. Precision medicine approaches can yield similar models that can be applied on a more individualized basis, and are an area of current research [268].

The additional real-time data collection and processing enabled by mHealth and IoMT devices will make it increasingly possible to quantify human beings such that the practice of evidence-based medicine can be individualized. Human quantification and data integration will present healthcare providers with additional tools to scientifically determine which patients are the most similar to their own, and thus perform "real-time" epidemiological research to decide on the best treatment using N-of-1 trials [224]. With the aforementioned tools, healthcare providers may be able to identify smaller segments of the population that need particular treatments with higher confidence. When data is integrated on large scales (e.g., the entire population of the United States), over long periods of time, it could potentially become practical to evaluate endpoints of interest (e.g., heart attack) rather than surrogate endpoints (e.g., blood cholesterol levels) to more effectively evaluate treatments. This practice of integrating large sources of genetic (e.g., DNA sequencing), behavioral, and environmental data for developing precise personalized treatment plans is known as precision medicine.

Inclusion of heterogeneous big data sources has the potential to have a transformative effect on evidence-based medical practice and allow for improved healthcare delivery [176]. However, this must be preceded by studies of the efficacy and sustainability of precision medicine interventions [176]. Enabling precision medicine to advance evidence-based medical practice is therefore a significant motivating factor for the deployment of IoT and cloud integration in healthcare.

9.4 Health Data Ownership in IoT and the Cloud

Technological advances within the healthcare industry (e.g., mHealth) have created an unprecedented amount of user-generated, health-related data [199,

257]. Data ownership and the implications on personal and data privacy from third parties attempting to connect to these devices in order to access, capture, analyze, and share this data remains an under-explored area from a policy and regulatory perspective [199, 237]. Thus, there is a clear need for transparent regulations and requirements for health-related data ownership and sharing.

9.4.1 IoT Data Ownership Challenges

Although ownership and protection of health data is an obvious concern within the healthcare field, the use of IoT-based health devices makes the issue more complex. In particular, some of the factors that further complicate establishing data ownership include the unobtrusive nature of the IoT device and its portability, and users' mobility, patterns, and preferences. Further, a vast majority of IoT devices feature unconventional user interfaces, which increases the difficulty of performing a large number of tasks (e.g., user consent and authentication). Given the diverse nature of IoT user interfaces, there currently is no "one size fits all" solution. Developing unique solutions for each IoT device further complicates the endeavor of developing standardized data ownership regulations.

9.4.1.1 Consent for Data Capture

A multitude of IoT-based health devices are continually capturing data from the outside environment. Due to the nature of continuous data collection, the data resulting from this process has the potential to include data sourced from nonconsensual collection, i.e., data collected from individuals without authorization or informed consent, or data collected from individuals unaware of the data collection. In these scenarios, it is difficult to designate the data owner: should the data belong to the owner of the IoT device, or should it belong to the individual whose data is being captured? While traditional IoT devices, such as smart doorbells, are in use, the owner of the device is often responsible for ensuring that any captured data is not violating privacy laws, irrespective of whether or not they are operating the device. Depending on these local laws, the owner might need to obtain authorized or informed consent from all individuals before they are captured by the device.

However, the ability to obtain authorized or informed consent through an IoT device is especially challenging due to its inherent characteristics (e.g., ubiquity, transparency). As an example, placing a physical sign regarding the data collection policies of an IoT device could easily go unnoticed, thereby not constituting authorization or informed consent from the recorded individual [101]. Whether or not an individual or user were to observe a physical sign, notice, or warning, the unique interfaces used in a vast number of IoT devices may prevent the individual from providing authorized consent. For example, audio interfaces used by voice assistants (e.g., Siri, Google Assistant) may not have a visual interface for the user to provide consent in a trivial manner. The ubiquity of IoT devices makes this process impractical since a user may need to authenticate and provide consent in a location with a large number of devices. Furthermore, multiple requests for consent have the potential to create user fatigue, thereby making user consent invalid and impractical [101].

In a healthcare setting, several of the challenges associated with obtaining authorized and informed consent of data collection from IoT devices can be resolved as part of the registration process for new patients. Similarly, in clinical research, researchers can obtain informed consent for all data collected by any IoT device used within a study as part of the standard informed consent process approved by their Institutional Review Board (IRB). In addition, IoT devices that collect clinically-relevant data outside of healthcare settings should be designed to only collect identifiable data from their consenting subject. Enabling real-time, on-device data processing may help to overcome this challenge. As an example, the Apple Watch provides a Noise app that performs local audio processing to enable users to understand the sound levels in environments that could negatively impact hearing. The Noise app performs local processing without recording audio content and thus does not currently require consent from inadvertently recorded bystanders in areas where it is available [14]. Additional frameworks for obtaining consent from IoT devices where these methods are impractical have been proposed, such as implementing informed consent through gateway devices– either directly from users or indirectly through a centralized registry [101]. For example, The Privacy Coach [76] scans RFID tags within IoT devices to compare the device's specifications to the user's privacy preferences.

9.4.1.2 Verifying Data Ownership: Local Identity Management and Authentication

Many IoT devices used in healthcare are primarily used as sensor devices to collect data. Since the data collected from these devices is being used in an increasing number of clinical decision-making processes, the sensor data can be manipulated in an adversarial manner to change clinical practice. In a potential future clinical situation without a human in the loop, these data integrity issues can lead to significant security gaps, as studied in the field of adversarial machine learning [171]. Even with a human in the loop, healthcare providers will rely on IoT-collected data to make clinical decisions that may be impacted by false data. This data integrity issue requires mechanisms for local user identity management and authentication in healthcare regardless of whether or not the device can provide direct access to clinical data. In practice, these adversaries could be third parties looking to cause harm, or patients looking to alter data for personal gain, such as a patient working to manipulate his data in order to be prescribed a controlled substance.

The distinct interfaces of IoT devices can make local authentication challenging. Even if alternative devices can be used to authenticate IoT devices through proximity, it can create enormous user burden and fatigue with ubiquitous IoT deployment. Developers of IoT devices that collect healthcare-relevant data must therefore balance the need for authentication with user burden to promote device compliance and limit security risks.

In addition to IoT devices that produce clinically-relevant data, some devices may consume produced data and provide feedback to users. Secure authentication is increasingly vital for these devices because of their potential to leak personally identifiable clinical data. In particular, these devices could reveal personally identifiable information (PII) that is traditionally respected as sensitive both within and outside of healthcare environments.

In order to assist in securing patient information, several IoT device authentication schemes have been proposed. Although these schemes are not widely accepted, they support the non-traditional interfaces of many IoT devices and could be applied to health devices. These authentication mechanisms vary in their overhead and burden to the user, and therefore may need to be considered on a device-by-device basis. These authentication mechanisms may also need to be used in conjunction with other mechanisms (either as an additional or alternative factor) to meet specific device requirements.

A non-comprehensive overview of several authentication mechanisms relevant to healthcare is listed below:

- **Proximity-based Authentication:** Several IoT device authentication approaches rely on the device's proximity to other user-owned devices, often acting as a physical authenticator. Relying on the prevalence of mobile devices such as smartphones, these mechanisms can require varying degrees of interaction with the user. At the simple extreme, proximity-based authentication might involve automatically authenticating devices within a specific range of proximity. On the more stringent extreme, when user attention is deemed necessary, proximity-based authentication mechanisms can require specific user action. Move2Auth [394] is an example of an interactive authentication scheme, wherein it requires users to perform specific hand gestures with a smartphone near the IoT device in order to authenticate. Because of their potential security vulnerabilities, proximity-based authentication schemes must be carefully evaluated before adoption in healthcare environments. Although elliptic curve cryptography (ECC)-based radio-frequency identification (RFID) IoT authentication schemes have been deployed in healthcare environments, all implementations might not have security requirements that are suitable for healthcare deployment [160]. Proximity-based and other physical authenticators may also be used to de-authenticate after proximity or physical contact has ended.

- **Biological Authentication:** Other IoT devices, especially those that already incorporate biometric sensors, may rely on biological authentication. Some biological factors that are used for authentication include fingerprints, face, heartbeat, iris, or voice [166]. Similar to proximity-based authenticators, some biological authentication mechanisms, in combination with other biological or non-biological factors, can also be used to de-authenticate after the biological factor changes or is removed. For example, IoT fitness devices can use heartbeat or capacitive sensors to detect when the device is removed and thus should be de-authenticated [368]. Because these factors themselves may be considered personal data, storing them directly on devices for authentication is highly discouraged. Instead, the configuration of IoT devices for biological authentication should utilize raw sensor data to train a mathematical model such that the original personal data cannot be reconstructed. Apple's TouchID technology used in select iPhones, iPads, and computers utilizes a similar approach [1].

- **Proxy-Based Authentication:** Proxy-based authentication relies on a secure channel between a specific proxy and the IoT device [86]. In this approach, a proxy (e.g., a clinician) would verify the user's identity and authenticate the device to the user. This proxy access could be granted to the clinician for a specific device to limit the possibility of this privilege being misused. Generally, when proxy-based authentication is required, devices should not de-authenticate with other factors in order to decrease burden on both the users and the proxies (often healthcare providers).

- **Behavioral Authentication:** The less commonly deployed strategy of behavioral authentication relies on non-biological behavior data to identify a user [286]. By relying on behavioral data, the user authentication process can often be implicit, with limited user burden. Although challenging, the development of novel secure behavioral identification techniques would have payoffs that could significantly improve the usability, and, therefore, the adoption of IoT device authentication. Current implementations have been developed that identify habits based on user data such as phone calls and locations, but are currently only suitable as secondary factors within multifactor authentication architectures [179, 328].

9.4.2   Healthcare Data Ownership

Much of the data currently used by healthcare providers is stored within health information technology systems. This data often takes the form of an electronic medical record (EMR), electronic health record (EHR), or personal health record (PHR). While these terms are often used interchangeably, there are some significant differences to these terms with regards to health data ownership. EMRs are traditionally owned by a single office or organization and are mainly a digital version of a traditional paper medical record [15]. EHRs are similar to EMRs in that they are owned by a healthcare organization, yet differ in the context that EHRs are designed to enable sharing with providers across healthcare organizations [15]. In contrast, PHRs defer ownership to the patient who collects and stores information across healthcare systems [4]. While EHRs are in widespread use today in the United States, personally owned records may become more widespread as patients begin sharing data collected by their own IoT devices with healthcare providers. These record systems are examined in further detail in the remainder of this subsection.

9.4.2.1 Electronic Health Record (EHR)

EHRs have achieved widespread adoption within the United States with 84% of hospital [292] and 53.9% of office-based physicians [11] adopting a basic EHR.[1] Adoption of fully functioning systems with additional features is lower. While the ideal EHR would allow for complete data federation across all healthcare providers, this is far from the case in many current isolated deployments. While there is limited work estimating EHR fragmentation, several studies have explored the extent of incomplete health data in EHR systems [70, 240], which can lead to patient harm such as medication errors [61].

The EHR is a healthcare organization-owned system in which patients have limited ownership of their own data. Although regulation, such as the Health Information Technology for Economic and Clinical Health (HITECH) Act in the United States [66], provides incentive for allowing electronic patient access to EHRs, some data remains unavailable. These restrictions can be beneficial, as direct access to test results has been shown to lead to anxiety and increased rates of patient visits [305]. However, allowing direct patient access has also been linked to increased patient engagement and is highly valued among patients [339, 361].

Although electronic health data is not widely used in research, their use in the clinical decision making process is increasing [167], and patients may be willing to share such data for research purposes [192]. With universal interoperability, the EHR can theoretically enable research leading to significant public-health benefits when fully adopted. For example, data trends can be used to

1 increase accuracy in influenza strain predictions for vaccinations,

2 evaluate treatment effectiveness in specific sub-populations, and

3 identify emerging drug resistance.

However, attempts at EHR interoperability face familiar barriers such as missing data [203]. In scenarios where complete interoperability has not been achieved, the EHR remains a healthcare organization-owned record. This can lead to a lack of efficiency but could also have harmful patient effects, such as redundant imaging [210], as patients visit healthcare providers using different EHR systems. Finally, there is prevailing belief that patients should have a right to accept the consequences to access and manage their own data [361], regardless of the potential risks [305]. This belief is related to the ethical principle of patient autonomy, and accepted practice of informed consent.

9.4.2.2 Personal Health Record

The PHR is a patient-centered form of medical record in which data is stored on a patient-owned portable device, or cloud service that the patient is able to

---

[1]A basic EHR, as defined by DesRoches et al. [110], includes support for patient demographics, patient problem lists, medication lists, clinical notes, prescription order entry, viewing laboratory results, and viewing imaging results.

access. This ensures that a patient's personal medical records are always available to them. When the patient moves from their primary care provider, to an urgent care clinic, to a medical testing facility, they are able to bring their record with them to be populated regardless of the record system used by the practice.

The PHR can also solve many of the data federation concerns associated with EHRs, as the patient is responsible for maintaining a single record which can be accessed by all care providers. Two well-known PHR systems were Google Health [232, 233] (closed in 2011) and Microsoft HealthVault [10] (closed in 2019), both of which originated from the Personal Internetworked Notary and Guardian (PING) or Indivo systems [12, 247, 332]. These PHR systems provided users with a single portal through which they could access their health information by linking with existing EHRs via interoperability standards such as the Continuity of Care Record [132]. Beyond that, these systems provided functionality for fine-grained data sharing as well as integration with personal health devices and health apps [10]. Unfortunately, the universal PHR systems have waned in favor of fragmentation, reminiscent of the current state of EHR [52]. In this fragmented model, users are connected to healthcare provider-owned EHRs as well as siloed commercial PHR systems such as the ones provided by pharmacies or fitness trackers. This fragmentation has occurred in context of a trend toward patient-driven self-care, including quantified self-tracking (e.g., PatientsLikeMe, 23andMe, Fitbit) [350], resulting in many, often data type-specific, PHR systems. Beyond the inconvenience of maintaining several record systems, the integration of these systems could limit patient harm by decreasing the prevalence of incomplete medical records [70].

However, the tide may be turning back toward a universal PHR system as tools such as Apple Health continue to gain in popularity and achieve high user satisfaction [7, 103, 283]. Apple Health is significantly different from Google Health and Microsoft HealthVault, in that it is a product of the smartphone era with the interface and data localized to an owner's device rather than an Internet portal. The obvious consequence of this is privacy: by retaining health data encrypted locally on a user's device, a user does not need to be as concerned about data misuse or the compromise as with an Internet-hosted portal. However, by hosting the PHR locally on an owner's device, Apple Health is also less able to provide two-way data transport: the app is able to collect data from healthcare providers, rather than currently providing data to providers. While the two-way data transport does not solve the problem of incomplete EHRs, Apple's CareKit [2], a developer framework for applications that allow patients to share health data with healthcare providers is touted as the potential solution. Apple Health utilizes an improved interoperability standard for accessing EHR data from participating healthcare providers [62], which is discussed in detail in Section 9.6.

9.4.2.3 Bridging Medical Data Ownership: Combining EHR and PHR

Currently, healthcare providers continue to maintain and collect patient data within EHRs, and patients collect personal data within (perhaps fragmented)

PHRs. It therefore follows that EHR-PHR interoperability will be an important step for mHealth and personal IoMT data utilization in healthcare. A simplified model of EHR-PHR interoperability that federates data sharing between both healthcare provider and personally owned IoMT devices is shown in Figure 9.1.

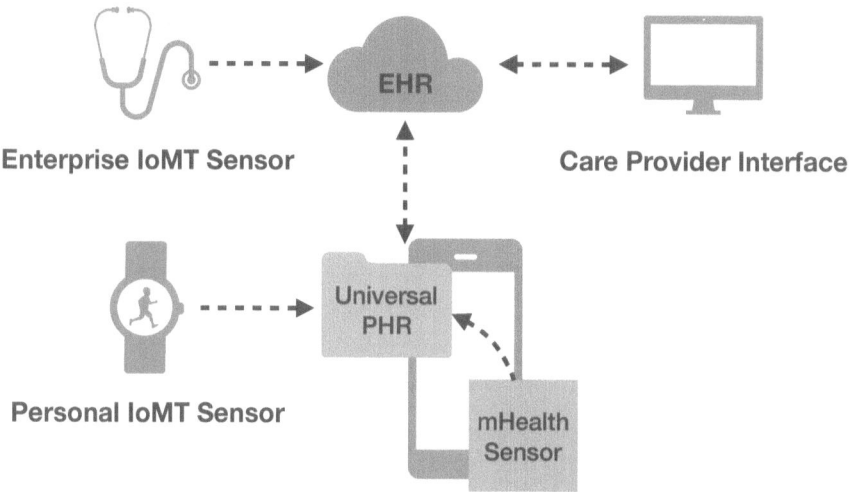

Figure 9.1: A simplified model of interoperability between an EHR and PHR system that can be used to integrate personal and enterprise IoMT devices. The PHR hosted locally on a user's smart phone can be replaced with a cloud-hosted PHR without modifying the model.

While this simplified model directly shows only a single EHR and universal PHR, additional EHRs could connect with the universal PHR via the same mechanism. By linking multiple EHR systems through a single universal patient-owned PHR, interoperability of both healthcare provider-owned and personally owned health records can, in effect, be achieved to prevent gaps in medical data. Additional architecture details for enabling health information exchange, such as cloud-hosted PHRs and 5G IoMT sensors, are discussed in Section 9.5.

In addition to mitigating EHR silos, combining EHR and PHR systems in this fashion also addresses a significant limitation associated with PHRs. As discussed previously, EHRs provide repositories of aggregated patient data that can be mined for public health and precision medicine research. Unless PHRs are hosted together, the PHR may lack the ability to serve as a research data repository [369]. While many EHR systems share only basic amounts of data, some large EHR networks, such as Epic [6], maintain interoperability between their EHR systems by including medical data that may be derived from IoMT devices. The integration of PHRs with these large connected EHR systems can produce data that would allow for public health research. It should be noted that access to care from providers that are part of these EHR network-based data warehouses may be limited for some populations, such as those with unequal access to care. Until universal access to these systems is established, clinical use of this data must therefore account for this skew.

9.5   Enabling IoMT Information Sharing in Healthcare

This section examines the strategies necessary for the federation of IoMT data in healthcare. The section starts by focusing on strategies for the health information exchange of traditional medical records. The section then discusses several challenges with federating IoMT data via similar mechanisms as well as current solutions, where available.

9.5.1   Collecting Data from IoMT Devices

Logically, the first step in being able to use data from IoMT devices is to collect the data generated by the IoMT devices and sensors. Since the IoMT devices used for support operations in hospitals are entirely enterprise owned, we will focus on IoMT devices operating within their 'clinical' use case.

At a high level, the collection of healthcare data from IoMT devices, whether the devices are enterprise owned or personally owned, has broad similarities. In both cases, there are two main considerations that need to be addressed for effective data collection. First is managing the enormous amounts of data that IoT devices are capable of generating. Decisions need to be made on what data is collected and the frequency of data collection. Several aspects of personalized medicine depend on integrating vast amounts of collected medical data for clinical research. However, this need should be balanced with identifying the important pieces of data since the amount of data collected also directly ties into the network resources required to transmit the data to the cloud or an alternate data storage location. Data thinning techniques should be applied to retain only essential data, so as to help reduce overhead in transport and data processing needs at a later stage. Second, and more important, is security. Personal data transmitted by devices that monitor health must be secure to protect personal privacy.

However, the primary differentiation between enterprise IoT devices as opposed to personal healthcare devices and Mobile IoT sensors is the user. The average user is not knowledgeable enough to make decisions on the magnitude of determining the collection frequency for personally owned IoT devices. Users also tend to be more at-risk for security exploits than enterprises practicing good software hygiene.

9.5.2 Traditional Health Record Information Exchange for Information Federation

Health information exchange is the sharing of patient-level electronic health information for assessment, cost reduction, and quality improvement in healthcare [369]. The HITECH Act mandates a limited level of health information exchange to be eligible for incentive payments in the United States [369]. Traditional information formats for medical record information exchange are discussed in Section 9.6.2. While healthcare provider relationships can enable exchange without significant technological access control requirements, additional considerations must be made when sharing data between patients and healthcare providers. The methodologies for traditional medical record information exchange can vary across emergency and non-emergency situations, and provide insights for the integration of personal IoMT and mHealth data into EHR systems [344].

9.5.2.1 Regulating Provider Access to PHR Data

Multiple fine-grained access control methods for PHR data have been described to provide secure information sharing of health data. Such an approach would allow for patients to have control over which users are able to access specific information contained within the patient's encrypted PHR. In order to enforce such control over their health records, patients would have the authority to generate and provide decryption keys based on the information they wish to provide to the receiving party. This method of access control would enable the secure sharing of PHR data with authorized healthcare providers while protecting the patient's personal data from unauthorized parties. However, many of these fine-grained access control methods result in high overhead costs when applied to scenarios involving multiple users, and thus impacting system usability. A fine-grained access control framework for PHR data with reduced overhead has been proposed by Li et al. [220]. This approach involves users generating their own sets of attribute-based encryption (ABE) keys. To account for the linearity of ABE encryption, the system is divided into multiple domains which are associated with various user subsets.

9.5.2.2 Providing Emergency Data Access

Although access control of medical records can be achieved through the previously discussed methods, in the case of required emergency access to data contained within the PHR and EHR, personal health data may become available without prior authorization. Currently, emergency data access to protected records within a single EMR system often follows a "break the glass" procedure. This procedure involves a healthcare provider self-granting access to a patient's medical record and protected health information (PHI) that can be utilized in the event of an emergency. Each instantiation of the "break the glass" procedure is documented and is later audited and reviewed to ensure that the patient's medical record and PHI were accessed under justifiable circumstances. Because an auditing, review, and accountability process exists under the "break the glass" procedure, it is not clear how this approach can be implemented for the sharing of health records across organizations using separate EMR systems or for granting access to PHRs in the case of emergencies.

The current practice of utilizing the "break the glass" procedure in emergency circumstances presents the risk for providing unnecessary access to patient information. One method of differentiating between users who should be granted emergency access to patient data and users who should be denied access integrates both Role Based Access Control (RBAC) and Experience Based Access Management (EBAM) strategies [398]. In order to test the effectiveness of this approach, the resulting algorithm, "Roll-Up", was applied to log data collected from Northwestern Memorial Hospital Center. Results from this case study indicate that a combination of RBAC and EBAM strategies is able to predict the conceptual position of a user requesting access to a patient's EMR data with 82.3% accuracy [398].

In the case of Li et al.'s proposed fine-grained access control framework, the issue of handling the security risks associated with providing data access during an emergency is handled using decryption keys [220]. This trapdoor method involves the patient selecting which parts of their PHR data they wish to be accessible in advance of a health emergency. The patient is able to delegate access of this data to the emergency department by providing a decryption key for each part of the pre-selected PHR data. These decryption keys would be stored within the emergency department's database of patient information. If an emergency occurs, a staff member would be able to query the database and obtain the patient's decryption keys from the emergency department. Once the patient's medical condition has returned to normal, the patient's PHR system could then compute re-keys for their PHR data and submit this update to the emergency department for future use. Although naturally supported by the framework proposed by Li et al., it remains unclear if this "break the glass" method would be able to scale and work across multiple hospital locations, given that the patient must be able to provide decryption keys to a particular emergency department in advance of any emergency incident. In order for this emergency data access method to be used across hospitals, the decryption keys provided by the patient would have to be stored in a centralized database, causing a host of other security and scalability issues.

Digital Rights Management (DRM) schemes can also be considered as a way to secure PHR and EHR data from unauthorized insider access. Kunzi et al. propose a data-centric model for the protection of health records in which encrypted health data is able to be accessed in an emergency with use of an emergency license [194]. Under such circumstances, an emergency key is issued in order to decrypt the patient's health data. Similar to the "breaking the glass" protocol, emergency access is documented and audited to ensure appropriate record access. However, in order to prevent against system compromise, a compromised emergency key will have a limited effect on the system. Additionally, the system design of this approach also ensures the dependability of the system while operating offline.

9.5.3   Ensuring Data Integrity from IoMT Sensors

Unlike traditional EHR-integrated sensors, the collection of data from mHealth and IoMT devices involves connections to large numbers of devices outside of the control of the healthcare organization. These devices may integrate via local gateways, connect directly to the EHR, or connect to cloud-hosted data warehouses. Regardless of the connection mechanism, the data collected from these devices will integrate with a health record system (likely an EHR or PHR). With potentially numerous connected devices, it is important to maintain a device inventory and validate that unaltered sensor data is being transmitted and received.

A simple mechanism to mitigate integrity risks for direct sensor device transmissions would be to utilize cryptographic encryption and signing, with validation performed based on public keys stored within a centralized device inventory. In this model, an mHealth or IoMT device would periodically be registered with the health record system, in which the device generates a key pair and shares the public key with the system's sensor device inventory. The mHealth or IoMT sensor would also receive a public key to encrypt the data sent to the health record system. The sensor could then encrypt and sign traffic to the health record system, which could be decrypted and then validated. The health record system would then tag the data source for each piece of received data. This method would help to ensure only data from valid, inventoried sensors is shared with the health record system, and provide a basic method for tracking the sources of received sensor data.

9.5.4   Privately Replicating and Sharing Large Datasets

In the EHR-PHR interoperability mechanism discussed previously, data is replicated and stored across numerous health record systems. As collected data sources become larger, such as when dealing with genome sequences, replicating data in its entirety across several cloud systems becomes impractical. A more efficient solution would be for record systems to use a "link" to a single instance of the data. The record systems which do not store the data in its entirety could initially download the data and compute any summary statistics necessary to store locally, before deleting the data. These calculated summary statistics might be used directly by the organization, such as common single nucleotide polymorphisms (or SNPs) from genomic data, or the number of steps traveled from motion data. Future calculations or analysis could be performed by simply downloading the file from the "link" again, or utilizing an application programming interface (API) provided by the file host.

This solution for enabling the sharing of large datasets does not address the ownership of the large file that is linked to by other sources. Naively, it could be proposed that the owner of the connected device that provides the data must also host these files. For example, if a patient received a genetic test from 23andMe, then 23andMe would be responsible for hosting the results indefinitely. While it is clear how this would work with tests from 23andMe or imaging from a healthcare provider, certain other scenarios aren't quite so straightforward. For example, how would motion data from personally owned devices be stored? What would happen if the company hosting the data went out of business? Would a user be responsible for paying to host their own data in order to receive the best care? Furthermore, when a single entity is responsible for hosting a data file, ensuring redundancy and availability can be prohibitively expensive.

An alternate solution is for these data files to be hosted using a consensus mechanism, in which small pieces of data are stored in a distributed fashion across several source hosts. These data pieces can be stored redundantly and in fault-tolerant fashion, such that if a single data host becomes unavailable, its shards would remain available from other hosts [59]. As a single file is replicated across additional sources, these sources would individually need to store less of the overall data. These data hosts can include EHR providers, commercial data providers or device manufacturers (e.g., 23andMe or Fitbit), and user-owned cloud storage that may be reimbursed by insurance providers or governments. Some published mechanisms for achieving this distributed data storage method would be suitable for this application in healthcare. The Security-Aware Efficient Distributed Storage (SA-EDS) model proposed by Li et al. requires data packets to be retrieved from a set of cloud storage providers before yielding the original data [219]. Similarly, Shafagh et al. propose a blockchain-based mechanism for secure distributed storage and sharing of time series data [325] that can also be modified to include genomic or other large non-time series data.

9.5.5   Maintaining Consensus in Large-Scale Federated Systems

Not all IoMT data is large enough to require efficient distributed storage. For smaller data sets that can be replicated across several medical records, there is an opportunity for maintaining consensus. Today, much of this burden falls on the patient. For example, although vaccination records are shared between healthcare providers, it is likely to be the patient who would catch a healthcare provider mistakenly administering a vaccination that they have already received. Often, these small mistakes may go unchecked, but, when caught, may lead to changes in treatment. For example, in a 2004 study of a computerized medication reconciliation tool, physicians changed the discharge orders for 94% of patients when discrepancies were identified by nursing staff [308].

Traditional software-only consensus solutions are inadequate in such situations as patients may report or present different information to different providers. For example, a patient might not take a medication prescribed by one physician and might not report (intentionally or unintentionally) taking the medication to another. In situations such as these, having the medical records retain a history of change would be a welcome feature. This would allow medication reconciliation software to automatically detect the discrepancy and allow healthcare providers to clarify information with the patient to update the record, which will retain the complete history of the reconciliation.

In medical records, retaining complete history is a necessary step for medical record reliability as allowing for the free deletion of medical records can allow patients to significantly affect the behavior of healthcare providers [381]. To ensure record reliability, changes to medical record values should not be allowed in EHR systems. In patient-controlled PHR systems, updated records should be assigned a new identifier such that they are shared with EHR systems as a new value. While some laws may locally require that patients are able to delete medical data stored in EHR systems [381], these deletions do not need to propagate– thus requiring patients to request their records be deleted across all EHR providers. This limitation would help prevent against the compromise of medical records within a large-scale federated health record system.

9.5.6 Providing Emergency Access to Real-Time IoMT Data

The collection and access of real-time IoMT data currently presents a potential challenge in the event of an emergency. IoMT devices are able to collect a variety of health data from users, including biometric data such as a user's heart rate, physical activity, and sleep cycle. Due to the high sampling rates of IoMT health monitoring services and applications, personal health monitoring devices have the ability to produce large sets of health data for each user. In the case that emergency access to this data is required for medical treatment, the volume of available patient data could result in finding the discernible 'signal' within this big data noise to be a challenging problem requiring the retrieval, sorting, and selection of relevant data.

Among these challenges is the issue of ensuring that the real-time data retrieved by the emergency department is the most up-to-date data collected from the patient's IoMT device. Currently, these devices may not sync to provide updates to a patient's EHR on a timely basis. Thus, the IoMT data retrieved under emergency circumstances may not be an accurate representation of the patient's current (or near-current) state of health. In order to obtain access to the patient's real-time IoMT data, the Emergency Department would likely have to be able to physically access the patient's personal IoMT collection device. This alternative access method may not be feasible in emergency situations as it relies on the assumption that a patient is conscious and able to give consent to allow for the emergency department to access to their IoMT device. Even if consent for emergency access to a personal device is granted, such access poses the risk of creating additional patient privacy concerns.

A potential path forward is for IoMT devices to enhance data availability by having the user select an interval where the device would update the user's real-time health data and offer a selected subset to emergency departments. This data update interval could include a range of data transfer periods for user selection. For example, when initially setting up their IoMT device, a user could decide to have their IoMT device automatically send their collected data to a selected emergency department(s) in hourly, daily, or weekly intervals. The emergency departments could then choose the interval in which previously received, and now out-of-date, data is discarded.

9.6   Achieving Heterogeneous Data Interoperability

Data interoperability in healthcare is a challenging problem even without the added complication of incorporating IoMT devices as data sources. While this interoperability is simpler in some countries which utilize nationally standardized EHR systems, interoperability challenges can still exist between jurisdictions. Outside of these areas, EHR fragmentation is exacerbated by enterprise-hosted deployments of proprietary and custom software that makes interoperability a significant challenge, as discussed earlier in this chapter. While interoperability standards do exist, these standards are often a subset of the data stored within a single EHR system. The integration of personal IoMT devices into this system only adds to the problem, requiring more comprehensive, flexible, and centralized integration standards.

This section will provide a general architecture for interoperability, and then discuss this current state of health record interoperability standards. Finally, it will describe potential alternative approaches for the interoperability of health data that can more directly enable IoT-cloud information sharing in healthcare.

9.6.1   Interoperability Architecture Overview

Compilers used in software development have a challenging interoperability task– to convert human-understandable source code to executable machine code that can be run on multiple target machines. At a high level, this process is broken into a front-end and a back-end, with the front-end yielding an intermediate representation of the source code, and the back-end converting that intermediate representation to machine code. It is this intermediate representation that prevents the need for a 1:1 match of compilers for every combination of languages and target computers, allowing the front-end to focus on the programming language and the back-end to make optimizations for the target machine.

Similarly, it would be problematic if promoting interoperability in healthcare required a specific tool to provide interoperability between each set of systems. The sheer magnitude of this task would make it impossible to enforce widespread interoperability via policy or encourage interoperability through market forces. Thus, IoT-cloud information sharing in healthcare would require a set of intermediate standards for interoperability. Such interoperability standards would need to be flexible enough to support the variety of data that can be made available through IoMT and mHealth devices.

While there are several more comprehensive models of interoperability, two commonly discussed variants of interoperability include syntactic interoperability and semantic interoperability [81]. Syntactic interoperability refers to the format and encoding used when data is transferred [81], such as the eXtensible Markup Language (XML) format used by Health Level 7's Clinical Document Architecture (CDA) [62]. The syntactic interoperability format would not help interpret the data, but provides the basic foundation structure necessary for interoperability. A set of modern, commonly accepted formats for syntactic interoperability, such as XML and JavaScript Object Notation (JSON), are already being used in similar applications and have additional widespread use outside of healthcare.

Semantic interoperability solidifies the meaning of data such that the meaning is not ambiguous as it is transferred, either between systems or humans [81]. This can be a challenge as health information can be represented both through different names or via a different structure in different systems. For example, one EMR system may represent a heart rate reading as a diagnostic event, while a PHR system might include heart rate information in a different structure designed for fitness data. Interoperability between these systems could be achieved by ensuring that they could both read and write to a comprehensive and flexible intermediate file format with its own representation and designation for heart rate data [356].

9.6.2   Current Interoperability Standards

Two commonly used examples of interoperability standards which dictate the formatting and exchange of health data are The American Society for Testing and Materials' Continuity of Care Record (CCR) and Health Level 7's Clinical Document Architecture (CDA). The purpose of both the CCR and CDA was to support healthcare data management and transfer between healthcare providers. Although the CCR and CDA were both created with the purpose of facilitating the collection and transfer of medical documentation between providers, and are specified in XML, they differ enough that they continue to co-exist.

The CCR was created specifically as a way for healthcare providers to collect patient information in an organized format that can easily be transferred between multiple care providers [132]. The resulting set of standards focuses on the patient's current state of health and other information, such as health insurance, care documentation, and practitioners. A key aspect of the CCR is its focus of presenting patient information in an easily human-readable format. A CCR document is divided into six sections including: an XML header, patient identifying information, patient financial and insurance information, the patient's health status, care documentation, and care recommendations. Although uncommon, IoMT providers are able to allow for the use of CCR data with their devices. Prior to the discontinuation of Google Health and Microsoft's HealthVault in 2011 and 2019, respectively, users of these services were able to integrate both personal health devices, health apps, and CCR data [3, 16].

As noted, the CDA was created to serve a similar purpose: to provide a method of standardization for the storage and transfer of healthcare documents. However, the CDA differs from the CCR on the aspect that the CDA was created to include different levels of machine readability in order to facilitate the transfer of documents between devices. In total, three distinct CDA levels exist: Level 1 is considered to be the most suitable formatting for older systems as it allows for an unstructured free text to be transferred between systems, Level 2 adds structure to the transferred documents by requiring the body of the document to be specified in XML, and Level 3 allows for the highest level of machine readability by requiring an encoded XML document [132]. Simply put, the interoperability of CDA specified documents increases with each subsequent CDA level.

9.6.3   Future Standards and Alternative Methods

Today, there are an increasing number of health-related mobile applications and IoMT devices being produced and used by consumers. Although these devices encourage users to take responsibility for the storage and use of their medical data, these personal devices may require both their users and healthcare providers to access brand-specific applications in order to retrieve health data. In addition to personal IoMT and mHealth devices, prescribed connected medical devices (e.g., remote cardiac monitoring portfolio from Abbott) also utilize web-based portals for healthcare provider access [9]. While some manufacturers provide health record integration solutions, such as Abbott's EHRDirect export [9], these integrations might not be feasible for some healthcare providers. This use of custom interfaces creates additional work for clinicians who wish to access patient data for use in treatment processes, serves as a barrier to data analysis tools, and emphasizes the growing need for updating the current interoperability standards.

To account for the rapid advances of the mHealth and IoMT device industry, interoperability standards need to be revised to allow for a greater degree of flexibility. As discussed in previous sections, the current interoperability standards do not allow for the easy addition of new healthcare data collected from personal devices. Future alterations to current standards could allow for mHealth and IoMT devices to automatically add recently collected data to a patient's medical record. This approach could introduce a standard way for both current and new device companies to provide updated user data to healthcare professionals without the use of custom applications for data access and retrieval. In addition, allowing for the device to update a patient's record automatically would remove the responsibility of providing a personal device's collected data to emergency departments from the patient. Such revisions to current interoperability standards could include methodologies that focus on incorporating a flexible editing and update process, or could take a more iterative approach.

Altering current interoperability standards to permit editing and updating of processes could allow for data to be added from mHealth and IoMT devices without the use of explicit standard updates being released. In this case, a newly released device would be able to add the user's data to their health record in the form of a device-specific field, which could then be updated with the user's most recent data in the future. These fields would ideally follow a standard format and could include information about the device itself in addition to the user's data such as: device manufacturer, device name, and date and time of last data update. Previously, larger companies such as Google and Microsoft have been able to provide integration of their devices with commonly used health record standards [3, 16]. An increase in flexibility which allows a way for mHealth and IoMT devices to add to and update a user's EHR in a standardized way has the potential to provide the same level of data sharing support to health device companies regardless of reputation, popularity, and size.

Similarly, the development of a universal, accepted set of standards would enable additional semantic interoperability of data for new IoMT and mHealth devices, and enable integration with EHRs. Lopez and Blobel have defined a development framework that could serve as a starting point for the design of such a standard relying on the Rational Unified Process (RUP) framework [235]. As such a semantic standard would be required to be comprehensive; an iterative framework similar to RUP may be able to allow for expansion during the development process. Despite their intended comprehensiveness, semantic interoperability models developed during these processes should be flexible to allow for the integration of new data types that may be similar to data already designed to be incorporated into the model. An example of such flexibility might be allowing a new fitness measurement recorded by a novel IoMT device to be recorded along with other fitness measurements in the interoperability format, along with a name and description of the previously unknown measurement type. Eventually, if this device and new fitness measurement become popular, this data can be added to future iterations of the standardized format. Reconciling data stored in this way with expanded standardized formats can be achieved using thesauri or word embedding, similar to the MetaNet technique developed for metadata domains [173].

9.7   Challenges & Opportunities

Precision medicine, as made possible by the advent and proliferation of IoMT devices and widely available genomic data sharing, holds great promise. However, success in being able to provide individualized evidence-based medical care is dependent on several key issues being addressed. Primary amongst these are:

1. Secure and reliable integration of data from clinical and personal health devices between enterprise IoT networks, private cloud networks, and personal connected health devices.

2. Since research regarding efficacy of mHealth interventions and outcomes is limited, continued initiatives to study the effectiveness of mHealth is essential to determining whether health-related IoT devices are engaging and providing actual value to users, rather than simply collecting data.

3. Settling ownership of data generated from IoMT devices is paramount to getting buy-in from customers wary of privacy violations.

4. A wide variety of IoMT devices makes achieving interoperability between these sensors to facilitate heterogeneous sharing of relevant IoT-derived health data between patients, healthcare providers, payers, and other authorized parties absolutely essential.

Chapter 10

Cloud Computing Based Intelligent Healthcare System

Sunita Pattanayak

Department of Computer Science and Engineering, University at Buffalo, State University of New York (SUNY)

The healthcare industry nowadays is so much more than just health of its patients. The need for a reliable, scalable and cost effective IT infrastructure to support growing demand for clinical, administrative as well as financial functions is leading to fast adoption of cloud technology in the healthcare organizations. Coupled with machine learning technology, it can be turned into an intelligent healthcare system. Machine learning is the next level of evolution in automation where machines learn from various data, eliminating the need for human intervention. It has the ability to process information already present beyond the capability of a human mind and, then, reliably analyze these data in improving decisions about patient diagnoses and finding more efficient treatment options. Combining these technologies, we get the intelligent cloud. It can learn from the vast amount of data already stored in the cloud to analyze difficult situations, make predictions and suggest efficient and convenient healthcare solutions. This can benefit in advancement of both the fields. As one of the IBM article states,

"Digital transformation has become an ongoing process rather than a one-time goal, with market-attuned companies continually on the hunt for the next big technology shift that gives them a competitive advantage. That next big shift is the fusion of artificial intelligence and cloud computing, which promises to be both a source of innovation and a means to accelerate change." The intelligent cloud is not to replace the doctors or the hospitals; it will rather assist the current healthcare technology to decide the effective ways of treatment. But what if it behaves differently and gives us wrong results? Can we afford such risks in healthcare? Here comes the role of interpretable machine learning. It deals with building trust in a machine learning model. The system being equipped with such advanced technologies, i.e., cloud computing and machine learning, also needs to be reliable, and interpretable machine learning provides us with trustworthy models. One of the important concerns in field of medical science is handling early detection and prediction of brain tumor in humans. For achieving this, detection and effective classification of types of tumors is necessary. The conventional method for the above issue includes getting images through MRI and then inspection by radiologists to identify the specific characteristics of the images produced. With such methods, handling large data and the requirement for reproducing any specific type of tumor are highly impractical, which is why there is a need for an intelligent cloud to get desired results.

10.1   Introduction

Cloud computing technologies are being widely used in the healthcare industry in numerous ways. International Data Corporation (IDC) estimated healthcare providers will account for 48% of total spending on industry cloud in 2018-2019. With growing challenges in healthcare, requirement for reliable, flexible or scalable hardware infrastructure has been continuously increasing and cloud computing is a one stop solution to all these requirements. Be it the healthcare apps, the medical equipment, the hospitals, the pharmaceutical companies, the insurance companies, the universities, or the healthcare platforms, all these companies take advantage of the cloud for various needs.

The data that flows in such industries is very crucial as it includes a patient's personal information as well as their medical histories. Protecting those data as well as making efficient use of these data cannot be achieved through cloud computing alone. Machine learning or data analytics play a very important role here in creating intelligent systems that will help humans achieve efficient healthcare solutions. Thus, cloud computing and machine learning can be used together, known as intelligent cloud, to discover new and effective ways of treatment. Not only in treatment, but also all the novel services mentioned in the previous paragraph can use this advanced technology to increase the ease of usage of such applications.

Even though security is provided by different cloud providers like Amazon Web Services (AWS), Azure, etc., trusting the predictions or solutions provided by a system is also required. Interpretable machine learning can help achieve the same by building trust in such kinds of intelligent cloud systems. In healthcare, it is very important to have reliable and trustworthy models that can help in creating not only innovative solutions but also the right solutions. The solutions that we get in such industries will have direct impact on the well-being of a person. This fact is threatening enough for us to make sure the solution is absolutely secure and trustworthy. For example, we will be discussing about handling early detection and prediction of brain tumor in humans and how we can achieve that, through the combination of these growing technologies.

For effective treatment of brain tumor, it is very important to detect and classify the types of tumors. The conventional method for the above issue includes getting images through Magnetic Resonance Imaging (MRI) and then inspection by radiologists is necessary to identify the specific characteristics of the images produced. With such methods, handling large data and the requirement for reproducing any specific type of tumor is highly impractical which is why there is a need to take aid of mathematical or computational models to handle such classifications.

10.2   Building an intelligent healthcare system

A huge number of factors come into picture when we are trying to build a product that will have more than one advanced technology, and understanding the need for such a system is very important. I will be discussing few advantages of building an intelligent healthcare system in the following paragraphs:

1. **Scalability**: With growing demand for healthcare needs and the dynamic nature of the needs in this industry, scalability is a major concern here. Due to the flexible nature of cloud computing technology, providing a scalable infrastructure will not be an issue.

2. **Reliability**: Considering the sensitivity of the data flowing in this industry, it is important to have a reliable system. Cloud computing provides a reliable platform as it ensures minimal data loss.

3. **Cost efficient research**: The intelligent cloud takes care of the data analysis and prediction from the data stored by use of its machine learning technologies. Also, the cloud provides low cost platform resulting in great research results at a fairly cheaper price.

4. **Dynamic Data**: Today, the hospitals or the healthcare applications use real time data to monitor patient's information and provide solutions. The intelligent healthcare system can help deal with such kind of data.

5. **Communication**: Healthcare industry consists of hospitals, universities, insurance companies, pharmaceutical companies, and research companies. With intelligent systems there is effective and faster communication between such modules in the industry.

6. **Security**: The data has to be secure which is ensured by the cloud computing platform. Also, the interpretable machine learning technology ensures safe use of such data and efficient as well as correct predictions or solutions.

Even though there are numerous advantages and use of such a system, there are various challenges as well in creating such a system. There has been a lot of work done in combining AI with cloud and creating a powerful system to be used across fields including health industry. But, segregating the right data, understanding the model that is giving us solutions and choosing the right solution requires a lot of time and effort. Interpretable machine learning is comparatively very new even though very useful. There are not enough models or enough work done in healthcare that can help us create a better model or trust one for the crucial problems existing in healthcare. As mentioned in [25], the future of interpretability in healthcare, there are still a large number of questions unaddressed in the area of interpretable models; but looking at the various solutions being discussed, it is a great area of research and can have unimaginable benefits in this industry.

10.3   Early detection and prediction of brain tumor using Intelligent Cloud

There are various kinds of machine learning as well as deep learning models that can be used for the classification of brain tumors. Naive Bayes classifier is one of the simple machine learning models as proposed by authors of [314] for classifying any set of data into specific types of known classes. It is one of the base models that we have used for classification to understand how accurately the model performs and if it can be relied on for such classification tasks. Convolution neural networks (CNN) have a very large learning capacity which make strong and mostly correct assumptions about the nature of images. It is also well known that the success of these networks largely depends on how much bigger a dataset is and how well trained the network is. But, the authors of [365] show that contrary to the belief that learning is necessary for building good image priors, a great deal of image statistics are captured by the structure of a convolutional image generator independent of learning. This motivated us to propose an interpretable CNN model for brain tumor classification to achieve effective classification.

An interpretable CNN model helps us in building a trust in the model. We achieve interpretability by inverting the inner layers of CNN and trying to restore back the original image with the tumor region as proposed by authors of [241]. This method allows us to analyze the features learned in the inner layer of the network and understand why a specific classification is made. Instead of the network depending totally on data for its training, we try to understand the mechanism behind the layers of the network architecture. By using the above mentioned approaches, we perform classification and compare the efficiency of each model. We also try to understand the advantages of using an interpretable model structure instead of a black box model.

10.3.1   Classification using different models

For better results and simpler training, Convolutional Neural Networks (CNN) models can be used for the classification task. A typical CNN model consists of an input layer, multiple hidden layers and an output layer. The hidden layers of a CNN typically consist of convolutional layers, pooling layers, fully connected layers and normalization layers. These layers are responsible for learning features of the image and use these characteristics for proper classification. Various authors have proposed interpretable CNN models as mentioned in [241]; more details of the architecture used are discussed in the experiment section.

Python can be used to implement the above described models. A Python package, Keras, makes implementation of all the layers involved in the twin architecture fairly simple and easy. Keras is a high-level neural network API, written in Python with tensorflow base for this project.

10.3.2   Image Inversion

The classification model used in Medical Science has to be trustworthy as the data is highly sensitive. To develop the trust in the model Deep learning model is interpreted, i.e., what the inner layers are learning about the input image is represented. This section introduces the algorithm to compute an approximate inverse of the input image. This is formulated as the problem of finding an image representation best matches the representation of input. Formally, given a representation function $\Phi : R^{HxWxC} \rightarrow Rd$ and a representation $\Phi 0 = \Phi(x0)$ to be inverted, reconstruction finds the image $x \in R^{HxWxC}$ that minimizes the objective:

$$x^* = \underset{x \in \mathbb{R}^{HxWxC}}{\arg \min}\, l(\Phi(x),\, \Phi_0) + \lambda R(x)) \tag{10.1}$$

where l is loss that compares $\Phi(x)$ to the target one Φ_0 and $\mathbb{R}$ is the regulariser capturing the natural image prior. The output of the above equation results in an image whose representation resembles the representation of input image. There may not be a unique solution to the problem and a sample space of possible reconstructions characterizes the space of images that the representation deems to be equivalent, thus revealing its invariances.

Loss Function. There are many possible choices of loss functions l but for our algorithm, we use Euclidean distance which can be calculated as per Eq. (10.2).

$$l(\Phi(x), \Phi_0) = \| \Phi(x) - \Phi_0 \|^2 \tag{10.2}$$

In this implementation, a normalized version of loss is used. Loss is divided by $\|\Phi_0\|^2$ so that the dynamic range of loss is fixed and can be contained in $[0,1)$ interval, touching zero at the optimum.

Regularisers. Discriminatively trained representations may discard some low level features as these are usually not as interesting as high level tasks. But as these low level features are useful for visualization, they can be partially recovered by restricting the inversion to a subset of natural images X. This restriction requires modelling the set X and as a proxy appropriate image prior can be in the reconstruction. In the paper, two such image priors are used. The first one is simply the $\alpha-$ norm $R\alpha(x) = |x\|_\alpha^\alpha$, where x is a vector. Divergence can be avoided by choosing a large value of α so that range of images stays within the target interval. The second regulariser is Total Variation (TV) $R_{v^\beta}(x)$, encouraging images to consist of piecewise constant patches. The formula for TV norm used is as depicted in Eq. (10.3).

$$R_{v^\beta}(x) = \sum_{i,j} \left((x_{i,j+1} - x_{i,j})^2 + (x_{i+1,j} - x_{i,j})^2 \right)^{\beta/2} \tag{10.3}$$

To make the dynamic range of the regulariser(s) comparable, one can aim for solution x^* to have unitary form. This requirement is considered by objective $\| \Phi(\sigma x) - \Phi_0 \|^2 / \|\Phi_0\|^2 + R(x)$ where the scaling σ is the average Euclidean norm of natural images in a training set. Also, the multiplier $\lambda\alpha$ of the $\alpha-$ norm regulariser should be selected to encourage the reconstructed image σ to be contained in a natural range [-B, B] (for our implementation B=128). The final form of objective function is expressed as per Eq. (10.4).

$$\| \Phi(\sigma x) - \Phi_0 \|_2^2 / \|\Phi_0\|_2^2 + \lambda_\alpha R_\alpha(x) + \lambda_{v^\beta} R_{v^\beta}(x) \tag{10.4}$$

10.4 Experiments and Results

The brain tumor dataset used in this project contains 3064 T1-weighted contrast-enhanced images from 233 patients with three kinds of brain tumor: meningioma (708 slices), glioma (1426 slices), and pituitary tumor (930 slices). Figures 10.1–10.6 shows images of three kinds of brain and tumor shape respectively. The data is organized in matlab© data format (.mat file).

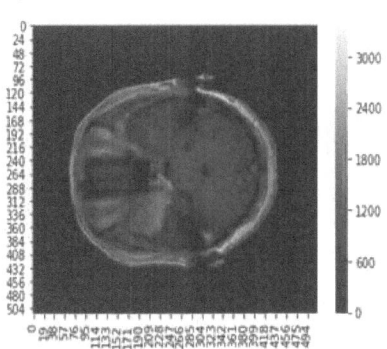

Figure 10.1: Type 1: Meningioma.

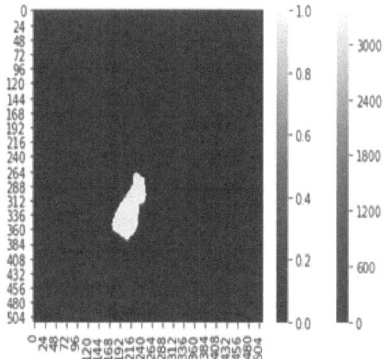

Figure 10.2: Type 1: tumor region.

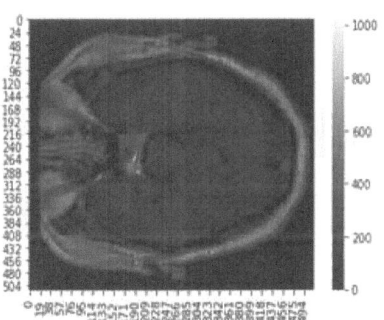

Figure 10.3: Type 2: glioma.

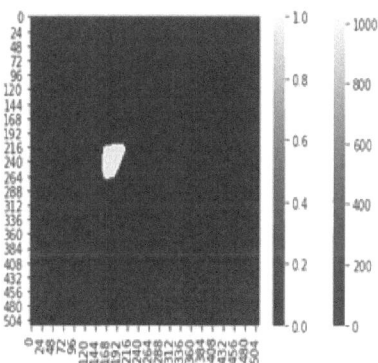

Figure 10.4: Type 2- tumor region.

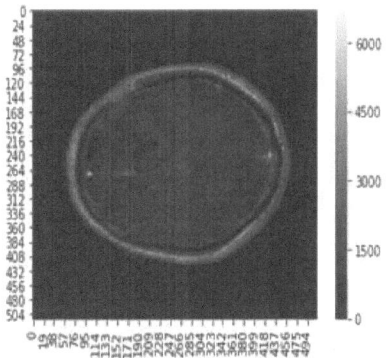

Figure 10.5: Type 3: Pituitary tumor

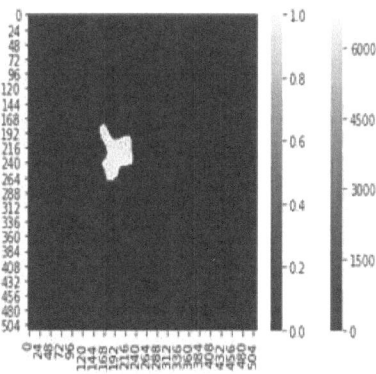

Figure 10.6: Type 3- tumor region.

Each file stores a struct containing the following fields for an image:

cjdata.label: 1 for meningioma, 2 for glioma, 3 for pituitary tumor.

cjdata.PID: patient ID.

cjdata.image: image data.

cjdata.tumorBorder: a vector storing the coordinates of discrete points on tumor border, for example, [x1, y1, x2, y2,...] in which x1, y1 are planar coordinates on tumor border. It was generated by manually delineating the tumor border. So we can use it to generate binary image of tumor mask.

cjdata.tumorMask: a binary image with 1s indicating tumor region.

Figure 10.7: Convolutional neural network layers.

10.4.1   Naive Bayes Classifier Model

For processing the data, we have used h5py python library to read the mat files, extract the image column from the 3096 images and store in a dataframe using Pandas Dataframe. We used 70 % of data to train the Naive Bayes model and tested the model on 30 % data. We have used the Gaussian Naive Bayes classification model from sklearn Python library to achieve the classification. We achieved a 89 % accuracy in this model. This model acts as a base model for our experiments and helps us classify the tumors using a simple, yet a learning model that uses available data.

10.4.2   CNN Model

We chose CNN model for its ease of training. Initially we started with just 2 convolutional layers and 1 dense layer with softmax activation function only to get 62.6 % accuracy. We made our model more efficient by increasing the inner layers. We have increased the number of convolutional layers, introduced pooling layers, added dropout and used 2 dense layers for the improved CNN architecture; the final architecture diagram is shown in the Figure 10.7. We achieved 73 % accuracy with the model.

We used softmax activation function so that we get the output as one of the 3 tumor classes. We use the category cross entropy as our loss function as we need a multiclass classification with probability between [0,1]. It helps us measure the performance of the model by comparing the predicted probability and actual labels.

For increasing the efficiency of the model, we used K fold cross validation method. The process included shuffling the dataset randomly, splitting the dataset into k groups. For each unique group, we took that group as a hold out or test data set and took the remaining groups as a training data set, then fitted a model on the training set and evaluated it on the test set. We then retained the evaluation score and summarized the skill of the model using the sample of model evaluation scores. We used 3 folds and achieved acc: 92.84 %, acc: 94.22 %, and acc: 97.06 % test accuracy in the 3 folds, respectively, with average accuracy 94.71 % (+/- 1.76 % standard deviation).

This helped us achieve a fair amount of accuracy in our classification method but for identifying back the tumor region, we needed a model that can predict the region or construct back the tumor images. We wanted to interpret the inner layers of CNN and the next method describes the method we used to achieve the same. The CNN model that we specifically used for this project can be found below. For better understanding of each layer, please refer to any convolutional neural networks journal as I have not talked about them here.

10.4.3   Image Inversion

After training the CNN model and validating its accuracy for classification, the weights of the model with the highest accuracy are saved. For image inversion the weights of the saved CNN classification model is used. We create a network with the same 2D convolution layers, activation layers and max pooling layers. The network does not contain dense and dropout layers. Also no batch normalization is needed in the network.

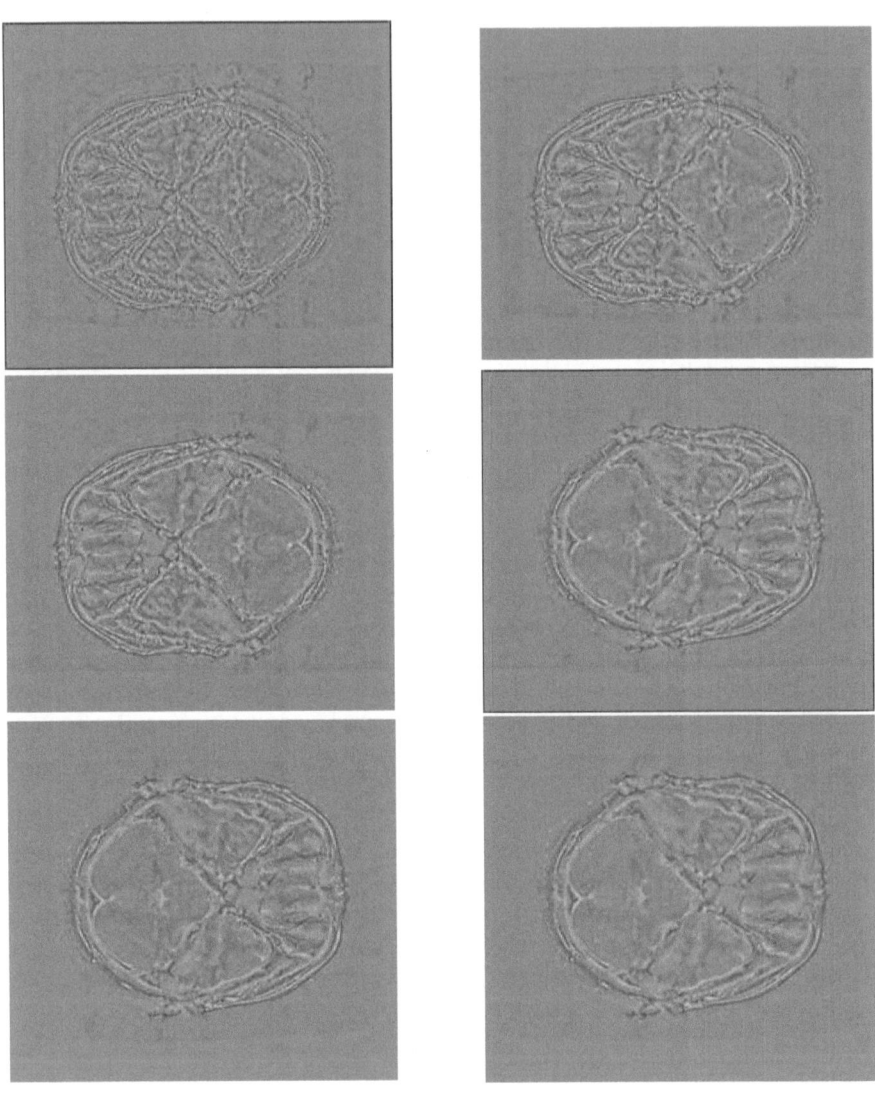

Figure 10.8: Images of features in last convolutional layer at every 100th
iteration for Type-1 tumor.

In image inversion, we interpret what the inner layers of CNN are learning. In our implementation, we are representing what is being learnt by the last convolutional layer (seventh layer of our seven convolutional layer network) which has 128 filters of kernel_size 2.

The images in Brain Tumor dataset are two dimensional. For our implementation, we reshaped the images to have four dimensions (batch_size, height, width, depth). Since a single image is given as input, the batch_size is 1. The input images are grayscale so depth of the image is 1. The width and height of the input image is 512.

The loss calculated in the network is the sum of Euclidean distance loss and total variation loss. The network uses Adam optimizer to train the network and minimize the loss. The network takes the input image and generates output at every 100th iteration with first generated image showing what last layer learnt when network is initialized with the weights of the classification model and raw image is given as input. For later iteration, the output of the previous iteration is used as input in the network. Figure 10.8 shows output at every 100th iteration depicting the minimization of loss in the process.

10.4.4   Summary and Discussion

This section proposed an interpretable convolution neural network model for classifying into various types of brain tumors and to invert shallow and deep representations in the inner layers of the model so that brain tumor regions can be constructed back for proper analysis and diagnosis even at an earlier stage. The image inversion method used with a simple classifying CNN gives us information learnt at each layer. This information helped us in reconstruction of the affected area which will help in earlier detection as well as prediction of a possible brain tumor occurrence. There are other CNN models which can be referred and enhanced to find better results. Such models can be found mentioned in [333, 402].

10.5   Research Challenges and possible solutions

While above solutions sound good in theory, it is extremely difficult to try and understand how various highly efficient existing machine learning models work under the hood. The complexity is added by the unknown as well. The introduction of interpretation in machine learning models is still very theoretical and have not seen much advancements in practice. Defining boundaries or scope to the interpretability in a model is also very difficult. Considering the industry in which this system is being considered, i.e., healthcare, the rise of cloud computing is huge but reliability on the other advanced technology, i.e., machine learning, is comparatively less and very complex. Only more research work and findings by using the already existing model can lead us to a more explainable model which in turn can utilize the existing cloud infrastructure and result in effective predictability.

Apart from the above challenge, most models are often biased to the type of data which are being learnt by the model. So learning the outputs of a particular model in various iterations or by studying the predictions again and again doesn't help in accurate interpretability. The interpretation to be really accurate and trustworthy might require identifying the blind spots in the processes and maybe segregating the right training data from the otherwise used data. In some cases, understanding the features which are taken into account in each layer in a deep learning model might also give a good insight on the interpretability. There are various ways in which it can be achieved and identifying that according to the problem we have and the solution we are trying to get is a major challenge as well.

10.6   Conclusion

The scope for research in creating an intelligent cloud system with interpretable machine learning is huge and with the growing needs and usage of these technologies in the healthcare industry, it is becoming the most efficient solution to cater such needs. As we could see in the previous sections, innovative as well as efficient solutions can now be achieved with such systems. Along with the scalable and reliable cloud infrastructure provided to process such huge amounts of data, these data can also be utilized towards innovations and improved solutions upgraded with new capabilities. The more we analyze the patterns that indicate beginning of a chronic disease, the more we predict more correctly such deadly diseases and provide early diagnosis to humans. More research and methodologies for understanding the interpretable methods and classification can be found discussed in [189, 94, 78, 228]. Thus, we can say collectively these modern technologies can bring in solutions we have not achieved until now.

10.7   Acknowledgement

I would like to thank Prof. Mingchen Gao, Bhumika Khatwani and other anonymous reviewers for the helpful and constructive discussions and feedback that helped in improving the quality of the chapter.

Chapter 11

IoT Cloud Network for Healthcare

Ashok Kumar Pradhan

SRM University, AP, Amaravati, India

E.Bhaskara Santhosh

SRM University, AP, Amaravati, India

Priyanka S

SRM University, AP, Amaravati, India

Due to rapid growth in information and communication technologies, the building of smart homes and cities becomes a reality. Smart homes and cities take the home and living experience to the next level. One of the major reasons for the development of smart homes and cities is to provide efficient and cost-effective healthcare facilities[100]. This chapter discusses different issues related to Internet of things (IoT) in Cloud network for smart Healthcare. The usage of IoT in healthcare has sharply increased across the industry and personal healthcare sectors. Remote monitoring and telemedicine are the main initiatives in IoT healthcare. More integrated approaches and benefits are sought with a role for the so-called Internet of Healthcare Things (IoHT) or Internet of Medical Things (IoMT). Remote health monitoring (RHM) is the main IoT use case in healthcare. Remote health monitoring and various other IoT use cases in healthcare are the main challenges in healthcare. In a health data context some data from medical devices and monitoring systems ultimately end up in Electronic Healthcare Records (EHR) Systems or in specific applications which are connected with them and send the data to labs, doctors, nurses and other parties involved. Finally, all the EHR are securely saved in the cloud so the doctors, patients and authorized third parties can access the information based on their requirements.

11.1 Introduction to modern health computing

Health is what makes or breaks an economy, which is why good healthcare is equal to a good economy. An efficient and reliable healthcare system is an important factor in establishing a good economy. Bad or poor health leads to inefficiency in work which results in various problems in economic balance. Provision of adequate healthcare facilities to each citizen of the country is the responsibility of government and is an important process in e-governance framework. In a country like India, where there is social and economic inequality, health is a major concern. People living in rural areas are not even aware of the basic and primary healthcare services. Villages constitute a huge part of the Indian economy, and they are not even getting the primary level medical facilities. According to the 2011 census[296], 68.84% of the total population of India is living in villages, but the condition of healthcare and medical services has not met the expectations and requirements. Rural areas face the problem of primary healthcare services shortage. Out of the primary healthcare sectors available, 8% do not have doctors or any medical staff, 39% do not have lab technicians, and 18% do not even have a pharmacist. These issues make the need for new and innovative technology an urgent and dire need to provide better healthcare services to rural India.

Internet of Things (IoT)[382] is an ever-growing technology that helps in connecting anything and everything over the internet. It aims at collecting data from various sources with high accuracy and less time. IoT enables collection and exchange of data which can be stored and used for analysis, measurement, as a reference by experts. It helps in making the world a better place and improves the quality of life. Information and communication technology can be effectively used to improve the healthcare system in rural areas[117]. Increased health awareness has led to the emergence of "self-care" and "healthcare advisor" disciplines. The benefits of a healthy lifestyle have fueled innovation that plays a key role in moving the point of care from the hospital or the physician's office to the patient's home. Delivering on this vision with the help of enabling technologies requires regularly capturing information related to a person's health, lifestyle, and other vital parameters and sharing it with caregivers. Smart patient healthcare monitoring systems provide better healthcare service by improving the availability and transparency of health data. However, it also poses serious threats to data security and privacy. The Fig.11.1 represents the fundamental features required for a novel smart healthcare system.

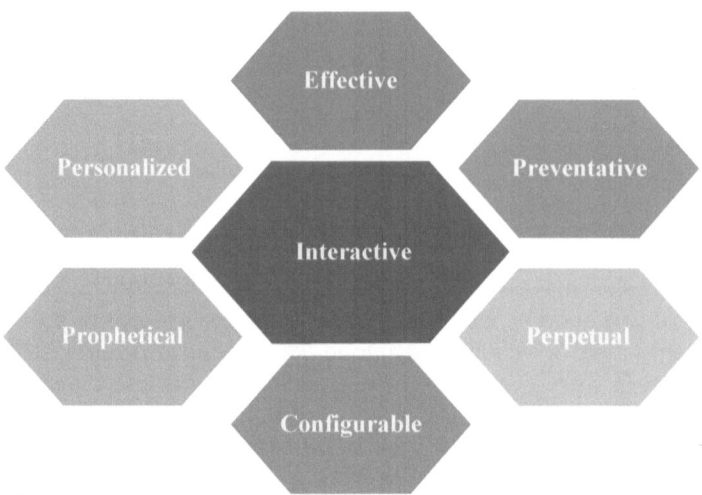

Figure 11.1: Novel Healthcare System Features

According to the latest survey by KaaIoT Technologies[8], spending on healthcare based IoT solutions will reach a strategic point of $1 trillion in the year 2025 by providing highly personalized, easily accessible and timely healthcare services for everyone. Several critical issues are considered to be important in smart healthcare, such as duration of monitoring, frequency of data collection and transmission, amount of data transmitted and nature of monitoring in terms of alerts, periodic or continuous. The following overview of requirements of patient monitoring shows the complexity, diversity, and somewhat contradictory nature of the requirements. At the same time, these healthcare systems need to improve reliability, efficiency and acceptability along with data security, privacy, and availability. As the healthcare industry has been increasingly getting more access to consumer data than ever before, health organizations are facing significant, but not insurmountable, challenges in managing, interpreting and protecting patient data. The lack of Electronic health record (EHR) integration is another barrier to overcome. The reliability and security of EHR data, interoperability, lack of proficient training in EHR management and lack of infrastructure are the hurdles among healthcare providers. The next tier of the problem lies among the populations that can benefit most from IoT like poor internet access among vulnerable populations, including the elderly, those with low education levels, lower-income populations, rural residents, and minorities. Let us now get to talk about overcoming these obstacles that the industry is currently struggling with.

11.2 Overcoming the challenges

Interoperability of disparate data sources needs to be addressed with IoT devices. As patient data comes in, health systems should have the infrastructure, resources, and processes in place to extract actionable insights from it that caregivers can use. Finally, as cyber-attacks are becoming a growing threat, health systems and their partners must ensure the protection of their networks by investing in viable features and capabilities. Since IoT devices capture and transmit data in real-time, the infrastructure to receive and process and store this data from millions of devices should be designed and built for scale. However, most IoT devices that report healthcare data suffer from a lack of data standards or protocols.

11.2.1 Security and privacy of patient data

Privacy and security concerns are slowing the progress for IoT to take over the sector and prove its potential. Healthcare is a highly regulated industry and requires everything to be secure and safe because patient information has to remain protected at any cost. However, security breaches still run rampant in many parts of the country. It is mandatory to meet the compliance requirements under the Health Insurance Portability and Accountability Act (HIPAA)[375]. When it comes to IoT based healthcare system, the compliance is also needed to focus on app developers, hosting service providers, cloud computing service providers and including subcontractors who involved in the healthcare space might have access to electronic patient health information (ePHI). Therefore, this industry is more focused on the virtualization of security right now than previously. Interestingly, many of the organizations that suffered breaches demonstrated a failure of proper controls over physical devices. Keeping physical devices under control is critical for gaining the confidence of patients in the use of IoT in healthcare. Providers, their associates, and vendor partners have a responsibility to ensure protocols are in place and their employees are trained to follow them.

11.2.2 Lack of uniformity among connected mobile devices

The problem is that there are no common standards or communication protocols to facilitate the process of aggregating information from them. The proliferation of connected health and activity devices that many of us now use makes it easy to see why a lack of interoperability is a huge obstacle to progress. Health systems should maintain at least some common standard for the type of devices kept in their facility. It should be done to facilitate a smoother transmission of data for quicker insights.

11.2.3   Vulnerable data transmissions

Ensuring connectivity is an important factor for IoT in healthcare. Data transmissions between devices, or between a device and the cloud, should be uninterrupted as well as speedy. Furthermore, they should have the capacity to host a great number of devices connected at the same time. Also, maintaining the quality and speed of transmissions is a key factor for IoT functioning. To overcome these challenges, the development of 5G technology is already underway. The other challenge is to make IoT sensors collect data even if there are troubles with the network. Also, an IoT system should be able to notify the user whenever a component is disconnected, so the physicians know at every moment what is going on.

11.2.4   Patient readiness

A non-technical, but significant factor is patient consent in adopting IoT. Patients are often confused about the introduction of new technology in a sector like healthcare and may be unwilling to take it. Physicians too may have their inhibitions about the same. Therefore, to overcome this challenge, patients need to aware of the potential benefits of IoT in healthcare. In a world that's slowly but steadily transitioning into a digitally-driven society, the applications of IoT are immense.

11.2.5   Awareness about IoTs

Understanding IoT from the consumer's perspective is not an easy task. As the uses for the IoTs are expanding and changing, there needs to be widespread awareness about them in the entire country. Furthermore, it requires constant push and promotion by healthcare authorities, physicians, care teams and patients to become successful.

11.2.6   Paralysis of Data Analysis

The overflow of massive amounts of data can lead to analysis paralysis. It means that it can be mind-boggling to go over every piece of information presented in the data. Extracting insights from data for analysis is the last stage of IoT implementation, and it has to be driven by cognitive technologies. Hospitals and health systems have a responsibility to ensure that the platform they opt for is capable of moulding according to their requirements. In smart cities, the IoT gathers a huge quantity of data and it can be processed by using automatic assessment systems. However, the increasing use of wireless transmission of health-related data raises the concern of data protection and authenticity. The medical data of an individual may be secured such that unauthorized access to the data could be denied. Only authorized healthcare staff may access the data to ensure the privacy of an individual's identity.

11.3 Cloud computing over the intelligent healthcare system

According to the American Association of Retired Persons (AARP)[245], 85% of senior citizens want to stay at home for treatment as long as the facilities are available. One of the concerns is that a large population around the world is aged 60 years or above. In the report of the United Nations on world population aging, which was published in 2015, it was mentioned that around 900 million people around the world are 60 years of age or above. This population will rise to 1402 million by 2030. Such a large population will occupy a significant portion of the health facilities in the hospitals. This situation can be avoided by building smart homes and cities, where automatic diagnosis systems will be a critical component. The automatic healthcare systems receive the data through the IoT and transmit it to central cloud for the evaluation. The cloud furthermore stores/mines the data and intelligently predicts patient's health status. It also provides feedback to patient through the computing device. The physician then can attend to the patient directly or send necessary precautions to patient via communicating device. Moreover, so far there is no automated medical server used in the field of healthcare as it requires large number of specialists to monitor the patient's health status data.

The cloud is equipped with complex and high speed computation modules. At the same time, the computation module accepts the new data from patient and processes it, then compares the obtained result with the previous results. If found suitable, the patient will receive feedback. In case the suitable record is not found, the appropriate physician will be informed through via phone call or SMS. The physician can get the information about patient from the cloud. Cloud storage capable of storing high volumes of varying data was also shown to be essential to a big data healthcare system by several previous works. If even a thousand people wore a single pulse sensor that communicated hourly with a cloud storage database via a low-power wide-area network (LPWAN), there would be 168,000 new data points per week. This number increases drastically as more people wear sensors connected to the cloud storage framework, and as more kinds of sensors are introduced. Using the big data that will rapidly form and continue to grow in cloud storage, machine learning algorithms can be implemented in the high-computing environment of the cloud.

These algorithms could be designed to mine through the large amount of data, identify previously unknown disease trends, and provide diagnostics, treatment plans, and much more. As we know the cloud storage is the most viable method for storing data. However, providing accessibility for healthcare professionals without compromising security is a key concern that should be addressed by researchers developing healthcare IoT systems. Machine learning offers the potential to identify trends in medical data that were previously unknown, provide treatment plans and diagnostics, and give recommendations to healthcare professionals that are specific to individual patients. Therefore, cloud storage architectures should be designed in a way to support the implementation of machine learning on big data sets.

11.4   IoT and smart health system paradigms

The following topics covers the broad area of IoT and smart health system paradigms.

11.4.1   History of IoT in healthcare

In the past decade, internet-connected devices have been introduced to patients in various forms. Whether data comes from fetal monitors, electrocardiograms, temperature monitors or blood glucose levels, tracking health information is vital for some patients, though many of these measures required follow-up interaction with a healthcare professional. Yet, the use of IoT devices has been instrumental in delivering more valuable, real-time data to doctors and reduces the need for direct patient-physician interaction. Early applications of IoT in healthcare are "smart beds," which detect when they are occupied and when a patient is attempting to get up. A smart bed can also adjust itself to ensure appropriate pressure and support are applied to the patient without the manual interaction of nurses. Another area where smart technology quickly became an asset in healthcare is when coupled with home medication dispensers. These dispensers automatically upload data to the cloud when medication isn't taken, or any other indicators for which the care team should be alerted.

11.4.2   Role of IoT in Healthcare

Turn Data Into Actions: Quantified health is going to be the future of healthcare because health that is measurable can be better improved. Therefore, it is wise to take advantage of quantified health technology. We also know that data affects system performance, and for that we depend on IoT devices.

- **Improve Patient Health:** What if the wearable device connected to a patient tells you when his heart-rate is going out of control. Moreover, updating personal health data of patients on the cloud and eliminating the need to feed it into the electronic medical records (EMRs), IoT ensures that every tiny detail is taken into consideration to make the most advantageous decisions for patients. Moreover, it can be used as a medical adherence and home monitoring tool.

- **Promote Preventive Care:** Prevention has become a primary area of focus as healthcare expenses are projected to grow unmanageable in the future.

The widespread access to real-time, high fidelity data on each individual's health will reform healthcare by helping people live healthier lives and prevent disease.

- **Enhance Patient Satisfaction & Engagement:** IoT can increase patient satisfaction by optimizing surgical workflow, e.g., informing about patient's discharge from surgery to their families. It can increase patient engagement by allowing patients to spend more time interacting with their physicians as it reduces the need for direct patient-physician interaction as devices connected to the internet are delivering valuable data.

- **Advance Care Management:** It can enable care teams to collect and connect millions of data points on personal fitness from wearables like heartrate, sleep, perspiration, temperature, and activity. Consequently, sensor-fed information can send out alerts to patients and caregivers in real-time so they get event-triggered messaging like alerts and triggers for elevated heartrate, etc. This will not just massively improve workflow optimization but, also, ensure that all care is managed from the comfort of home.

- **Advance Population Health Management:** IoT enables providers to integrate devices to observe the growth of wearables as data captured by the device will fill in the data that is otherwise missed out in electronic health record (EHR). Care teams can receive insight driven prioritization and use IoT for home monitoring of chronic diseases. This is another way that caregivers can make their presence felt in daily lives of the patients.

11.4.3   Challenges of IoT in healthcare

Internet of Things (IoT) technology implementations have raised numerous concerns around personal data privacy and security. While many of today's devices use secure methods to communicate information to the cloud, they could still be vulnerable to hackers. Beyond personal data being stolen and misused, IoT devices can be used for harm. For example, IoT in healthcare can be life-threatening if not properly secured.

Example 1: A 2012 episode of "Homeland" demonstrated a hack of a pacemaker inducing a heart attack. Later the vice president of company, Dick Cheney, subsequently asked the wireless capabilities of that pacemaker be disabled.

Example 2: In 2016, Johnson & Johnson warned one of its connected insulin pumps was susceptible to attack, potentially allowing patients to deliver unauthorized insulin injections.

Example 3: Then, in 2017, St. Jude released patches for its vulnerable remote monitoring system of implantable pacemakers and defibrillator devices. These are just a few of the medical IoT attacks that have made headlines.

To counter these risks, the U.S. Food and Drug Administration (FDA) has published numerous guidelines for establishing end-to-end security for connected medical devices, and regulators will likely continue to regulate connected devices used by patients. In late 2018, the FDA signed a memorandum of agreement with the Department of Homeland Security to implement a new medical device cybersecurity framework to be established by both agencies. It also issued a draft update to its pre-market guidance for connected healthcare device manufacturers in 2018 to ensure end-to-end security is built in during device design and development stages. The possible solutions are (1) end-to-end communication encryption, (2) embedded secure code implementation, (3) regular software updates, (4) mutual authentication, and (5) device identification.

11.4.4   Future of IoT in healthcare

In present day the IoT medical applications are available with multiple user-friendly configuration options along with simplified user interface, so, hospitals and healthcare sectors are no longer need to wait for training to deploy. Next generation devices are in the implementation or post-implementation stages. The future of IoT in healthcare is now. In fact, Aruba Networks predicted that 87% of healthcare organizations will be using IoT technology in their facilities by the end of 2019. Furthermore, it stated that 73% of applications of IoT in healthcare will be used for remote patient monitoring and maintenance, 50% for remote operation and control, and 47% for location-based services.

11.4.5   Patient-centered care

Healthcare is shifting its priorities from hospital centered services to patient centered services. The Institute of Medicine suggested an approach for improvement and "crossing the quality chasm" by outlining six aims for healthcare to be safe, effective, patient-centered, timely, efficient and equitable. Among the principles that had been proposed, the one that garnered most attention was the aim for healthcare to be "patient-centered by providing care that is respectful of and responsive to individual patient preferences, needs and values and ensuring that patient values guide all clinical decisions." Eight dimensions of patient-centered care as outlined by the Picker Institute include:

1. Respect for patient's values, preferences and expressed need;

2. Coordination and integration of care;

3. Information, communication and education;

4. Physical comfort;

5. Emotional support and alleviation of fear and anxiety;

6. Involvement of family and friends;

7. Transition and continuity;

8. Healthcare technology accessibility to every one.

The other beam of comfort lies in the information technology arena, whose epoch-making invention and development have been at the vanguard of human progress, in recent history. Today, technological advancements have made digital tools widely accessible and handy to the masses with approximately 46% of the world's population having access to the internet in 2016 and nearly 7.683 billion people having mobile cellular subscriptions in 2017[72]. Owing to this accessibility to the digital world, people have now become introduced to a boundless sphere of information, effortless communication, and endless opportunities by literally a click of the finger. Harnessing upon this massive penetration, technology has been deployed in healthcare which was advocated by the Institute of Medicine (IoM) as a vital means to accomplish the aforementioned six aims. Additionally, the World Health Organization (WHO) also resonated with the essential role of technology in realizing the 2030 sustainable development goals related to health. [82]. In the 1960s, advancements in communication technology and information and communication technology (ICT) opened doors to Telemedicine, which is literally means of "healing at a distance" and, according to the Institute of Medicine, it is defined as "the use of electronic information and communications technologies to provide and support healthcare when distance separates the participants."

Soon, it was recognized that the approach towards the remote provision of healthcare needed to encompass a more comprehensive outlook by incorporating non-physician related care, such as nursing, pharmacy and elements of public health education along with promotion of self-care. This broader scope of telemedicine was coined as Telehealth. The turn of the century witnessed a colossal upsurge of the internet and every sector including healthcare went on board to benefit from the fresh opportunities that now lay before them. This led to the rise of Electronic Health (eHealth) which is defined as "an emerging field in the intersection of medical informatics, public health, and business, referring to health services and information delivered or enhanced through the Internet and related technologies. In a broader sense, the term characterizes not only a technical development, but also a state-of-mind, a way of thinking, an attitude, and a commitment for networked, global thinking, to improve healthcare locally, regionally, and worldwide by using information and communication technology."

But what may be considered a game changer was the proliferation of mobile phones into the hands of a common man. Capitalizing upon this accessibility to technology, the healthcare sector found new ways to address the healthcare challenges facing them, heralding the rise of mobile health (mHealth). The mHealth is a subset of eHealth and provides medical and public health services and information via mobile technologies such as mobile phones and personal digital assistants (PDAs). The mHealth offers a means for healthcare professionals to keep their patients updated via reminders, alerts and health-related information. Multiple studies have focused on the role of mobile SMS as a means for impelling behavior change, self-efficacy and improving knowledge in areas such as sexual and reproductive health.

With the coming of the Internet of Things (IoT) into the picture, it became possible to create a network between different devices through software, sensors and network connectivity thereby enabling an exchange of data between them. This propelled the rise of Digital Health which is defined as "an improvement in the way healthcare provision is conceived and delivered by healthcare providers through the use of information and communication technologies to monitor and improve the wellbeing and health of patients and to empower patients in the management of their health and that of their families" and includes categories such as mHealth, health information technology (IT), wearable devices, tele-health and tele-medicine and personalized medicine.

In recent times, with the realization of bringing patient-centric values of patient engagement and empowerment to the forefront, adoption of the latest technology in healthcare is become more common. With this scenario a new socio-technical concept of "Connected Health" came into reality. Its aim is to make health and wellness services safe, effective, efficient and as a result enhance the quality of life and lower costs. Connected Health is an overarching model that includes all aspects of technology use in healthcare such as telemedicine, telehealth, mHealth, and eHealth. Furthermore it mirrors gaps between technologies for information sharing and connectedness together for proactive care and integrated healthcare services. Moreover, it has opened up a new vista in healthcare by digitally connecting clinicians to clinicians, patients to clinicians and patients to other patients. Hospital facilities too are progressing in parallel by utilizing technological innovations to enhance the care and safety of the patients during their stay at the hospital, for instance, by installing automation systems in the building to regulate temperature, ventilation, and fixing smart locks. Interconnected clinical information systems such as Laboratory Information Systems ensure smart patient care processes. Moreover, identification systems enable authentication and tracking of patients, staff, and hospital equipment.

11.4.6   Teleconsultation and Remote Patient monitoring

The ample opportunities for effective communication resulting from technological advancements have laid the groundwork to enable real-time consultations between health providers and patients separated by geographical distance, a process known as teleconsultation and thus bridging the communication gap between them. A more robust form of teleconsultation is remote patient monitoring (RPM) which deploys the latest IT tools to provide diagnostic and treatment services to the patients located in remote and rural areas. For instance, Alentejo, an underserved region in Portugal with regard to adequate and accessible healthcare, successfully initiated telemedicine and teleconsultation in 1998 as a means to improve healthcare and is now an essential part of service delivery there. Moreover, a systematic review highlighted the feasibility of telemedicine in the field of dentistry for remote screening, diagnosis, and consultation. Additionally, teleconsultations with the health provider have been found to enrich patient satisfaction due to improved outcomes, ease of use, low cost, better communication and reduced travel time. Furthermore, studies have underlined strong support in favor of telemedicine in the aspect of patient safety since it has been revealed that use of telemedicine for consultation brought down the number of medical errors in between clinical visits, besides playing a part in lowering medication errors. As excessive waiting time at the hospital continues to be a pressing problem faced by patients, the efficiency of e-consultations to provide convenient access to healthcare professionals may be considered.

11.4.7   Wearable sensors

Miniaturized, sensor-enabled wearable devices have made it plausible for patients with chronic diseases such as cardiovascular disease and diabetes to monitor their vital signs such as blood pressure and blood glucose level and thus indulge in self-care. It further allows the patient to transfer the data obtained to a healthcare professional using wireless technology. A review highlighted the feasibility of wearable devices in the promotion of physical activity and weight loss. Moreover, the data obtained from the wearable sensors alert the patient and the healthcare team regarding adverse events and prompts timely remedial action. The Fig.11.2 represents the wearable sensors in healthcare.

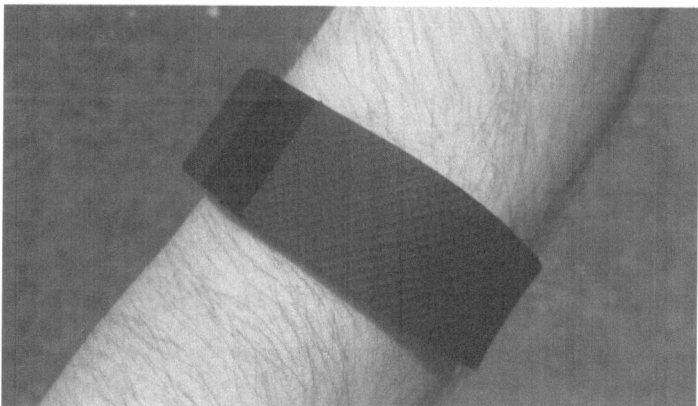

Source: Wuefab
Figure 11.2: Wearable sensors in Healthcare

11.4.8   Insideable devices

Unlike wearable sensors which usually remain in contact with the skin, ingestible sensors gauge the internal changes in homeostatic imbalance and offer novel means to diagnose and monitor the human body. An ingestible sensor has been approved to monitor medication compliance among patients with hypertension and heart failure. Another novel technique of monitoring is by way of implantable sensors which can be positioned below the skin and permits the measurement of vital signs, for example, Cardio MEMS is an implantable device which helps in continuously monitoring pulmonary artery pressure. A randomized clinical trial revealed a reduction in hospitalization of patients with chronic heart failure by 50% when their daily pressures were monitored [156].

11.4.9   Mobile apps

Smartphones with inbuilt health apps provide a unique opportunity for patient engagement by promoting, adopting and maintaining healthy behaviors. As of 2015, approximately 165,000 health-related apps are available and are broadly classified as "wellness management apps" [253] which assist in modifying behaviors related to lifestyle, diet, fitness, etc., and "health condition management" apps which facilitate dealing with disease conditions by providing information about the disease, access to care and medication reminders. Chronic conditions including mental health conditions, diabetes, cardiovascular diseases, nervous system disorders and musculoskeletal conditions are amongst the most common conditions focused on health condition management apps.

11.4.10   Electronic Medical Records (EMR)

EMRs that can store health and medical information of a patient in digital form have widely attracted physicians; for instance, in Canada, approximately 75% [250]of physicians have shifted to EMR use. Besides improving the communication between the healthcare team, it delivers them readable and organized information which reduces the risk of medical errors. The Fig.11.3 represents samples of EMR in healthcare.

11.4.11   Health portals

Aimed at bridging the communication gap between the patient and providers, portals are personal healthcare-related websites that allow the patients to communicate with their healthcare team through teleconsultations. Moreover, they permit access to lab test results, schedule appointments with the doctors and refill prescriptions. A systematic review of the effect of patient portals concluded that ten out of twenty-seven studies reported positive effects in terms of medication adherence, self-care practices, improved patient satisfaction and functional status.

11.4.12   Big data

As a result of the digitalization of medical and health records (EMRs) and data generated from wearable devices, a large and complex volume of data is being produced known as Big Data. This massive reservoir of information is now being put to use by assisting clinicians in providing an observational evidence base.

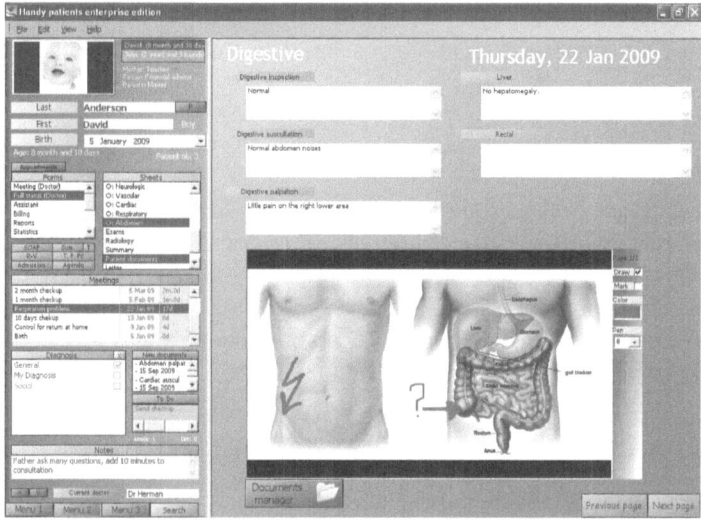

Source: en.wikipedia.org-electronic health record
Figure 11.3: Electronic Medical Records (EMR)

Big Data has also facilitated the opportunity to deliver personalized treatment by using analytics in assimilating genomics with EMR.

11.4.13 The human genome project

By far the most monumental scientific discovery in recent times was the unraveling of information regarding the structure, organization and function of the human genome undertaken by an international research collaboration known as the Human Genome Project. This project was an epitome of a partnership between biologists and technologists since the investigation into the genome applied computing technology extensively and these days, owing to further advances in biomedical technology, a sea change in the diagnosis and treatment of diseases is anticipated.

11.4.14 Personalized and precision medicine

As a result of advancements made through the Human Genome Project[111] in understanding a person's genetic makeup which determines their susceptibility to certain diseases, it is now possible to provide tailored therapies suitable for each patient, thereby making them safer and effective. Personalized medicine takes into account not just the genetic makeup of individuals but also their preferences, beliefs, attitudes, knowledge and social context. On the other hand, precision medicine utilizes patient centrism, engagement, digital health application, genomics, molecular technologies and data sharing in healthcare delivery.

11.4.15   3D Printing

3D printing refers to the "deposition of materials such as plastic, metal, ceramics, powders, liquids or living cells in layers to produce a 3D object." It is redesigning healthcare since it is now possible to recreate body parts such as personalized prosthetics. Remarkably, Spanish scientists have successfully launched a prototype for a 3D bioprinter that can create a fully functional human skin and can be transplanted to burn victims[270].

11.4.16   Artificial intelligence in healthcare

An exciting dimension to the digitalization of healthcare is the development of intelligent machines which exhibit cognitive actions analogous to human beings and are capable of conducting real-time analytics using algorithms. For instance, IBM Watson helps clinicians make decisions by using natural language capabilities, hypothesis generation, and evidence-based learning. This mechanism is particularly useful given the surge in Big Data and will assist in excavating information and aid the doctors in making a quick and precise diagnosis. The potential role of an artificial conversation agent or Chatbot, which uses speech or textual methods to conduct a conversation is being explored in healthcare to assist with behaviour change in diabetes and obesity management. Additionally, Babylon Health is a conversational health service provider which uses artificial intelligence to have consultations with doctors.

CASE STUDY 1

A real-time system for the monitoring of blood glucose levels in diabetic patients. This system requires patients to take blood-glucose readings at set intervals manually. It after that considers two kinds of blood glucose abnormalities. The first is abnormal blood glucose levels, and the second is a missed blood-glucose reading. The system then analyses the severity of the abnormality, and decides whom to notify, the patient themselves, caregivers and family members, or emergency healthcare providers such as doctors. This system is practical, and it is realizable.

CASE STUDY 2

A system aimed at detecting heart attacks was built using ready-made components and a custom antenna. An electrocardiogram (ECG) sensor is used to measure heart activity, which is processed by a microcontroller. This information is shared via Bluetooth to the user's smartphone, where the ECG data is further processed and is presented in a user application. The authors identify that developing heart attack prediction software would improve the system. Further improvements could be made by measuring the respiratory rate, which is known to aid in the prediction of a heart attack.

11.5 New Design and Performance of IoT cloud for Smart Healthcare and Monitor system

Cloud computing converges new technologies and existing ones to offer as services with all capabilities of a computing system to different kinds of users. These services can be accessed from anywhere with the help of an internet connection independent of their physical location. The Internet of Things (IoT) is a platform that enables the capturing of real-time information, facilitates examination and analysis of this information and provides a connected environment by sharing it with various stakeholders. Cloud and IoT are mutually dependent on each other. IoT can benefit from the virtually unlimited capabilities and resources of Cloud to compensate for its technological constraints (e.g., storage, processing, and energy). Cloud can benefit from IoT by extending its scope to deal with real things in the real world and for delivering a large number of new services in a distributed and dynamic manner. The new designs for healthcare and monitoring systems must possess the following characteristics.

1. **On-Demand Self Service:** IoT Cloud Computing is readily available to all when you need it. As Cloud Computing resources are a web-based service, it can be accessed without any help or permission from others. But the most primary need for establishing communication is by internet, because internet is everything in the world.

2. **Broad Network Access:** IoT Cloud Computing provides a lot of connectivity options. Cloud computing resources can be accessed via tablets, mobile devices and laptops with internet connection. This makes it easier for the user to easily access the devices that they mostly like. Without the help of IoT, cloud computing can't be accessed and function; that's why networks are most significant nowadays.

3. **Resource Pooling:** Resource Pooling means that it can be shared for those who know where resources address. Resource pooling will make people know the address that can be accessed anytime and anywhere as they want. It makes the user able to access what they want and when they have free time to access. In IoT context, an IP address can be easily assigned to every "thing" on the planet they want like computing IP address accessing.

4. **Rapid Elasticity:** In IoT Cloud computing with Rapid Elasticity you get what you need, because of its rapid elasticity nature. This cloud computing

provides the freedom to suit with what you need. You can easily and quickly edit your software features and add or remove user inside your cloud computing.

5. **Measured Service:** IoT Cloud Computing is a Measured Service in the meaning that you get what you pay for. This cloud computing will measure your usage about of their services such as storage, processing, bandwidth and active user accounts inside your cloud computing. The meter will increase as much as you utilize the resources. This system is known as Pay Per Use (PPU).

The healthcare industry is predominantly moving towards affordable, accessible and quality healthcare. All organizations are striving hard to build communication compatibility among the wide range of devices that have operated independently. IoT and adopt IoT driven systems and processes have the potential to model this kind of healthcare, which heavily relies on patient participation. This will subsequently improve the way health services are being delivered. IoT is here to stay, and will continue to evolve fast, leading to impactful and positive changes for all stakeholders in the healthcare industry.

There's no doubt that the Internet of Things (IoT) creates a new technological ecosystem in healthcare, helping organizations manage bottlenecks in care and bridging the gap between patients and providers. The list of applications for connected devices in hospitals keeps growing from real-time staff tracking across facilities to ensuring preventive equipment maintenance and 24/7 patient vitals monitoring in ER and resuscitation departments. But care delivery doesn't stop after discharge. A range of smart consumer-facing medical devices can ensure that the patient's health status will be kept under control between office visits. This will allow providers to grasp a multitude of subjective and objective symptoms, analyze them, build patterns and observe them in dynamics. Additionally, the patient becomes more informed and invested in their own health management, whether they have a chronic condition or not. With valuable insights about regular processes or sudden sensations, such as mood, sleep, temperature, pain, cough and more, patients gain a better understanding of how their own body works.

Experts acknowledge the growing role of IoT in healthcare, too. For instance, Deloitte anticipates the healthcare IoT market volume to reach $158.1B by 2022. However, with changes come challenges. As many connected medical devices gather patient vitals in chronic and acute cases, assuring their performance becomes the vendors' top priority. Any performance interruptions bear potential risks for patient safety and health outcomes, with varying severity depending on a particular device's intended purpose. Therefore, Quality Assurance (QA) specialists are under increased pressure when carrying out performance and compatibility testing of medical IoT devices. If the team is unaware of domain specifics and isn't ready to mitigate arising challenges, the resulting product may fall short of the high requirements for medical software or even fail FDA approval. The challenges in the IoT clotd smart healthcare system are discussed as follows.

11.5.1 Disruptions in Internet

When checking the performance of medical IoT software, testing specialists deal with the load, network bandwidth, latency, and other metrics both for mobile and web applications. While it is undesirable for a website to crash under an unexpected load surge, in healthcare IoT crashes are unacceptable, especially not for smart medical devices directly involved in patient care, such as a continuous patient monitoring system or a smart insulin pump. Still, force majeures do happen, and it is critical to prepare the product accordingly. When testing consumer-facing IoT devices, some QA teams need to take all the important variables into account. For example, use conditions can be highly diverse, especially when the connected device is designed for use in rural areas. From smart thermometers to blood coagulation meters and inhalers, the stability of the user's internet connection can twist the test results or even undermine them if the receiving specialists weren't aware of the actual network bandwidth. To try and compensate for such pitfalls, both developers and QA specialists have to ensure that service disruptions won't affect data transmission. For example, all sensitive PHI data can be backed up prior to transferring, so that each data package could be erased only after the sending attempt is successful. Additionally, performance engineers need to keep in mind that IoT devices don't operate with planned downtimes, so any patches and updates have to be integrated seamlessly into the device operation. It's highly recommended to test the process at least with the first patch in order to polish it and ensure that new updates won't disrupt user experience or tamper with sensitive data.

11.5.2 Diversity of Protocols

As connected medical devices are entering the mainstream, they are creating a whole new niche in digital care delivery. With this comes another performance testing challenge: devices frequently need new protocols to manage communication with the servers. The healthcare IoT ecosystem is already complex since it lacks standardization of IoT protocols currently in use[128]. While the Message Queuing Telemetry Transport (MQTT) protocol is one of the most common because it handles low bandwidth networks and low-memory devices, there are also HyperText Transfer Protocol (HTTP), Constrained Application Protocol (CoAP), Extensible Messaging and Presence Protocol (XMPP), Data Distribution Service (DDS), and many others. The diversity of available protocols and the devices they support puts an additional strain on performance engineers as they have to find a fitting load test tool covering this particular combination. But that's not the only difficulty in choosing the right toolset.

11.5.3   No Special Testing Tools Were Made for Healthcare Applications

There's no better way for QA specialists to hop into a clinician's shoes than deciding on the testing toolkit for a particular connected device. They need to pick the right instruments to arrive at a relevant diagnosis as well as treatment suggestions. Healthcare IoT architectures are complex and comprise unique combinations of devices, sensors, and actuators communicating with disparate systems via distinct protocols, generating and transferring enormous amounts of data. Not to forget that patient data is highly sensitive, so each transaction should be secure. To find a fitting load testing tool, performance engineers can break down the infrastructure in question and arrange the parameters according to their load testing priorities. For example, if both continuous data sharing and scaling to multiple devices are required, it's worth compromising only when there are thousands or millions of devices. Continuous data sharing, however, should be prioritized in any toolkit.

11.5.4   Difficulties in Performing Healthcare IoT Performance Testing

While testing the performance of a connected medical device might seem similar to that of a mobile app or a website, the complexity of IoT architectures creates new challenges for QA teams. These challenges can overwhelm even experienced specialists because they have to evaluate the system while always keeping in mind patient safety and data integrity. Here are a few suggestions on how to effectively deal with the challenges described above:

- Emphasize atypical use cases: whether you are evaluating a bedside monitor or an implantable smart device, users can and will get creative with the system.

- Don't try to find a perfect load testing product supporting all protocols within your healthcare IoT stack. Pick the tool with an SDK to start building tests right away.

- Narrow down your test scenarios by defining the most popular device/software version pairings to ensure more effective coverage. Then, test all new devices that communicate with the network and ensure that data will stay safe in case of internet network disruptions.

As mHealth [383] evolves, the growing adoption of machine-to-machine (M2M) technologies is helping bring greater automation to remote monitoring, as wireless technology is used to transfer real-time data about patient vital signs and conditions directly to medical staff. M2M technology can also help in the tracking of drugs and medical equipment and can enable better management of healthcare workflows. For patients, this means continued monitoring and treatment delivered in a way that is more convenient, less disruptive and that ultimately enables them to enjoy a better quality of life.

11.5.5 Mobile technology in revolution of Smart Healthcare

Creative use of new mobile and wearable health information and sensing technologies (mHealth) has the potential to reduce the cost of healthcare and improve well-being in numerous ways. These applications are being developed in a variety of domains, but rigorous research is needed to examine the potential, as well as the challenges, of utilizing mobile technologies to improve health outcomes. Health outcomes are defined as those events occurring as a result of an intervention. These may be measured clinically (physical examination, laboratory testing, imaging), self-reported, or observed (such as gait or movement fluctuations seen by a healthcare provider or caregiver). Some health outcomes require complex assessments to determine if they are present or absent. For example, some conditions, such as dementia, can be classified differently in studies depending on the country of the study population.

11.5.6 Financial challenges

An ageing population, an increase in chronic diseases and growing demand and mobile health services offer a way to improve the quality of care while reducing costs. In this changing environment, cloud computing can be an enabler of organizational transformation. Harnessing the power of the cloud, healthcare organizations can create dynamic infrastructures that improve operational effectiveness and dexterity by optimizing and accelerating IT resource and service delivery. Clinical and business boundaries can be erased by simplifying access to information, and connecting people as well as business functions across formerly siloed systems, while improving the economics of their IT infrastructure. Forward-thinking organizations are turning to more advanced technology that can take the information in systems and records and deliver it as cloud computing services. Cloud computing can provide a resilient technology infrastructure that delivers continuously available information-based services. When utilized effectively, cloud capabilities can enable enterprises to become more agile, reduce IT spending, as well as develop and deploy applications faster. Cloud computing can also help reduce energy costs, carbon emissions and the need to expand data centers[178].

Because cloud computing automates virtualization, the provisioning process is streamlined and of taking days to weeks of manual provisioning, an automated process can complete the task in minutes. IT infrastructure resources can be delivered more quickly, and the solutions sitting on that infrastructure can be globally available and scale dynamically. A recent IBM Institute for Business Value study[13], which surveyed 750 chief technology officer (CTOs), Chief information officer (CIOs) and other technology executives in 19 industries, including healthcare, highlights the strategic importance of the IT infrastructure. Over 70 percent of organizations recognized the important role IT plays in enabling competitive advantage and optimizing revenue and profit. Yet less than 10 percent of respondents reported that their existing IT infrastructure is fully prepared to address the proliferation of mobile devices, social media, data analytics and cloud computing. Despite the stated importance of IT infrastructure, only 22 percent of companies surveyed have a well-defined enterprise IT infrastructure strategy roadmap in place. This research highlights the challenges healthcare organizations face as they grapple with the new era of IT. The first step toward transforming the enterprise with cloud computing begins with the preparation of the IT infrastructure for more advanced cloud computing strategies. Complete transformation to the cloud is a journey not limited to infrastructure, though it is a fundamental business shift in how IT services are developed, financed and delivered. Cloud computing enables healthcare organizations to rapidly develop platform services in preparation for the digital health revolution and the growing importance of remote care services utilizing information from the Internet of Things (IoT).

11.5.7   SaaS helps improve delivery of Hospital services

A Canadian hospital needed to streamline clinical processes to improve patient operations and enterprise workflow, while simultaneously providing a mobile experience for clinicians. The hospital was able to reduce implementation time through IBM® BlueWorks Live[5], a business process management software as a service (SaaS) offering from IBM. Implementing the mobile process-management technology improved the hospital's throughput and patient experience. The outcome was a 15 percent improvement in staff productivity within three months. With the ability to support real-time analytics across data and organizational silos, and address management and cost challenges stemming from exponential growth of data and sprawling infrastructure footprints, cloud computing provides an effective approach for optimizing operations across the entire organization. As a tool for business model innovation, cloud computing can help healthcare organizations meet the imperatives of a transforming industry, to become more efficient, information-driven and patient-centric. Cloud services can help organizations to:

1. **Build sustainable healthcare systems:** Create an efficient, flexible organization that proactively manages requirements and opportunities to help overcome the operational challenges of controlling costs, improving efficiency, complying with regulations, optimizing resource utilization, and enabling better visibility across the infrastructure.

2. **Collaborate to improve care and outcomes:** Improve the quality and efficiency of care while cultivating patient centricity by overcoming challenges in the implementation of electronic medical records, promoting collaboration within and among care teams, and integrating secure, trusted information for analytics, evidence-based decision support and personalized care.

3. **Increase access to healthcare:** Reduce disparities in access and transform individuals into advocates for their own health by addressing challenges by

analyzing patient needs and behavior, adapting resources and delivery networks to meet these needs, anticipating demands, and delivering services to individual consumers and providers

Cloud technology connects hundreds of clinicians for those in need. A non-profit organization is using an IBM cloud-based social business solution that provides collaboration services to a global network of healthcare volunteers. With a focus on Haiti, the solution supports clinicians by giving them immediate access to critical data and information to help support the healthcare needs of the island's citizens. Using the IBM SmartCloud® [373]for Social Business solution to virtually connect medical workers and volunteers, those on the front lines taking care of patients are armed with an online medical knowledge system that includes treatment options, clinical pathways and best practices specific to the location.

11.5.8   The benefits of cloud computing

Cloud computing technologies are well suited for organizations looking for proactive ways to meet current healthcare industry challenges. Organizations can use cloud technologies to reveal valuable insights in their data and transform how they make decisions. Cloud solutions can virtually connect healthcare professionals around the globe to collaborate, respond more quickly, enable remote care and share best practices.

1. **Speed:** Cloud computing increases the speed of business innovation. Cloud technology supports the development of new applications more quickly than ever before using composable services from a marketplace of APIs. As a result, the organization can gain on-demand access to IT infrastructure resources including servers and storage networking and get feedback faster so the business can adjust accordingly. Healthcare organizations can respond more readily to the needs of the business, and the needs of the patients they serve, as well as their partners, suppliers and employees.

2. **Empowerment:** Cloud technology is helping healthcare organizations improve and re-engineer their business processes and workflows, as well as increase engagement and collaboration internally and across their enterprise. By freeing application developers to focus on expanding sophistication, rather than administrative and integration challenges, cloud technology helps IT professionals drive optimization and innovation instead of constantly building and maintaining their infrastructure.

3. **Economics:** Cloud technology can help improve a healthcare organization's economics with the ability to bring new capabilities to market ahead of the competition. Using a cloud infrastructure, an organization can buy the IT resources that it needs, when needed, enabling capital to be redeployed by shifting large, upfront expenses to variable expenses. The cloud can

also help an organization realize cost savings through automation and standardization.

11.5.9   Cloud security and regulatory compliance

The promise of business applications and IT solutions delivered through the cloud is compelling because it can provide new business capabilities on demand. However, it's important to note that there are significant security and compliance requirements which need to be addressed as part of cloud readiness and governance when using cloud services. These requirements are addressable through robust security, strategic planning and governance. In the US, the Health Insurance Portability and Accountability Act (HIPAA) and the Health Information Technology for Economic and Clinical Health Act (HITECH)[262] provisions indicate that healthcare organizations, covered entities (CE), and cloud service providers (CSPs)[387] each bear significant responsibility for regulatory compliance. Regulations regarding protected health data compliance vary by country and not all regulations are clearly defined. Companies looking to globalize business services through the cloud need to consider these variations in regulatory compliance as a significant element of any cloud strategy in terms of evaluating and managing risk across many countries.

Physicians are now able to monitor information about patients with chronic illnesses remotely. They are able to provide cost-effective home-based care to patients undergoing long-term treatment and can help elderly patients maintain independence, by creating safe and secure living environments. Greater access to smart mobile technologies, combined with more readily-available health information online, and interactive social media and mobile health services, are enabling patients and doctors to have a more engaged relationship. As mHealth evolves, the growing adoption of machine-to-machine (M2M) technologies is helping bring greater automation to remote monitoring, as wireless technology is used to transfer real-time data about patient vital signs and conditions directly to medical staff. M2M technology can also help in the tracking of drugs and medical equipment and can enable better management of healthcare workflows. While America leads the way, Europe and the developed Asia-Pacific are also at the forefront in embracing this technology.

11.5.10   Spend less money, serve more patients

With mobile access becoming increasingly prevalent worldwide and almost all developed markets having more than 100 percent penetration, mobile technology is well placed to support the delivery of healthcare. This can be done in a number of ways. With 53 percent of the Japanese population expected to be over the age of 65 by 2030, as well as 43 percent of Western Europe and 33 percent of North America, mHealth is helping to ease the burden of an ageing population in the global healthcare industry[256]. Mobile devices such as Global Positioning

System (GPS) locators, for instance, are being used to monitor the status of patients. The devices send real-time alerts in the event of an accident or sudden fall, meaning those who are less mobile or suffer from degenerative diseases such as dementia are able to stay at home and maintain a level of personal freedom, while receiving continuous care and protection.

As diseases once considered incurable, but now deemed treatable, place greater demand on the healthcare industry, mobile technology is reducing the impact on resources. Mobile devices enable clinicians to receive readings remotely and maintain up-to-date records on patients, resulting in less hospital visits despite more information ultimately being gathered on the patient. At any one time caregivers are also able to see whether patients are taking their medication correctly and can keep an eye on the condition of medical equipment to avoid breakdowns. While clinical trials support investment in new treatments and drugs, they can be time-consuming and costly. Vital sign monitors enable administrators to capture real-time patient data during clinical trials, speeding up regulatory evaluations, removing the need for manual, paper-based methods and facilitating faster decision-making.

It might be easy to assume that growing automation means that patients are taking an increasingly passive role in their own treatment programmes. However, often it is the combination of data collected from devices and data manually entered by patients that is most responsible for improving patient outcomes. It is the increasing collaboration between doctors and patients, facilitated by mobile communications, that is changing the face of healthcare delivery.

11.5.11 mHealth in action

- In South Africa global healthcare group Sanofi has launched a mobile-based Patient Support Program (PSP). Once a patient is diagnosed with diabetes, they can choose to receive SMS alerts to support them through their first six months of treatment.

- AstraZeneca is using mHealth services to improve global health outcomes for patients with cardiovascular conditions. Mobile and Internet-based services support a patient throughout his treatment, improving medication adherence and giving the patient the confidence to manage his condition more effectively. Access to personalised educational materials, as well as coaching and treatment, enables patients to manage their medication and lifestyle changes, while tracking their treatment progress.

11.5.11.1 IoMT Platforms

Almost every big provider of cloud, software, and infrastructure is trying to own the market in this growing sector. They are providing their services using IaaS, PaaS, and SaaS to develop and deploy healthcare specific services that provide value-add to an individual user of IoMT devices. Moreover, they are providing services to businesses, manufacturers, and hospitals to intelligently interpret data as well as defining actions for a specific user or a generalized pattern for a group of users. Most of these available platforms follow the same value chain as explained in the above section, but geared toward data gathering from devices, to data integration, to analytics, to trigger action for users or enterprise applications. The distinguishing factors across platforms are mainly around stability, scalability, the range of services they provide to support the deployed framework, support, and cost. Many of the IoT reference architectures needed to provide an end-to-end solution usually include the required components as mentioned.

1. **Endpoint IoMT Devices:** Modern or traditional health devices generating data towards cloud components.

2. **Network Gateways:** Internal or external devices that collect data (i.e., may have minimal to optimal capability to filter the collected data and provide connectivity to data centers through the Internet); devices may also have the capability to connect directly to data centers.

3. **Ingestion:** Could be an on-premise or off-premise data center, or a cloud component that gathers data from devices and provides it to other components for further processing and intelligence.

4. **Business Logic and Orchestration:** This could be an on-premise or off-premise data center or a cloud component. It usually builds with integrating data and applying defined rules and policies based on the specific platform application functions and SLAs. Offline device management and state maintenance may also become a part of this, or it can be a separate software component for maintaining device registry, state, device identity, authentication, and security.

5. **Analytics Software:** This too is usually an on-premise or off-premise data center or a cloud component that runs big data analysis on the data collected from the devices.

6. **Service Notification:** Most of the components inside an on-premise or off-premise data center or a cloud solution interact with this service for triggers and actions towards devices or an internal component of the cloud or an enterprise application.

7. **Storage:** An integral part of the solution to maintain collected data, configuration data, provisioning data, rules and policy data, operational data, control data, etc.

8. **Integration Bus:** Facilitates communication between different platforms as an application trying to access a platform's services and data.

9. **End User Applications:** IoMT platforms should enable the capability to develop and deploy applications (e.g., web applications, backend services), as well as reporting and monitoring applications to leverage the analytics performed on the stored or received stream of data from devices.

There could be many other software components that could be part of the solution, but they generally fall into one of the other categories mentioned above. Some vendors such as AWS, Azure, Qualcomm, and Intel provide IoT platforms that facilitate healthcare applications. This next section will explore each of these solutions.

11.5.11.2 Amazon Web Services IoT

Amazon Web Services (AWS) IoT can provide bi-directional and secured communication between IoMT devices and the AWS cloud. This is where healthcare companies can host their framework for specific requirements related to the healthcare domain they are dealing with, and can use AWS provided services to support integration, storage, and analytics needs other than the infrastructure provided by AWS cloud. According to AWS documentation, AWS IoT can support billions of devices and trillions of messages, as well as process and route those messages to both AWS endpoints and other devices reliably and securely. With AWS IoT, applications can keep track of and communicate with all devices, all the time, even when they are not connected. AWS IoT consists of the following components:

1. **Device Gateway:** This enables devices to communicate with AWS IoT securely.

2. **Message Broker:** A mechanism for publishing and subscribing to messaging among devices and AWS IoT applications; also integrates with services like Amazon Kinesis, Amazon Simple Queue Service, Amazon Simple Notification Service, etc.

3. **Rules Engine:** Message processing and integration with other AWS services such as Amazon S3, Amazon DynamoDB, AWS Lambda, etc.

4. **Security and Identity Service:** Provides security via device identity authentication, as well as secured communication with AWS IoT applications; it also secures data exchanges across other AWS services.

5. **Thing Registry:** Also referred to as the device registry for associated resources stored or used within IoT platform.

6. **Thing Shadow:** Also known as the device shadow and used to replicate the current device state-related data.

7. **Thing Shadows Service:** Logical representation of devices in AWS IoT to serve offline requests and connectivity in case of network communication unavailability.

Many healthcare-related companies are using the AWS IoT Platform.

- **Clinical Information Systems:** Calgary Scientific, Practice Fusion

- **Population Health and Analytics:** IMS Health, Philips, HC1

- **Health Administration:** Captricity, Infor, Pegasystems

- **Life Sciences Solutions:** DNA Nexus, Medidata, BioTeam, Synapse, Core Informatics

As far as compliance and security are concerned, several certifications, laws, regulations and compliances cover AWS. It is also compliant with Federal Information Processing Standards (FIPS), HIPAA, Health Information Technology for Economic and Clinical Health Act (HITECH), and Payment Card Industry Data Security Standard (PCI). Vendors using AWS IoT usually create an IoMT Application Server Framework specific to their unique healthcare domain. They deploy these frameworks under an Amazon Elastic Compute Cloud (EC2) infrastructure to handle the ingestion of traffic, apply specific handling, and use the rest of the AWS IoT services for the other handlings. For example, Phillips deployed its HealthSuite Digital Platform using the AWS infrastructure to make its devices interoperable and better manage the continuous stream of data generated. Phillips is also leveraging the AWS IoT platform for unified and custom communication protocols, secured mutual authentication, device "shadows" or virtual versions of devices, and a "rules engine" for managing specific types of data. Phillips is using almost all of the AWS services mentioned above.

11.5.11.3   Qualcomm Life

Qualcomm Life provides an E2E (end-to-end) solution for patients, hospitals, and medical professionals. Qualcomm launched its IoMT solution in 2015 via the 2net™ Hub and 2net Platform, a cloud-based platform that works with a variety of medical devices and other applications to help collect, transmit and store a patient's data. This data is communicated through a plug-and-play gateway that uses short-range radios (e.g., Bluetooth, WiFi). The below figure demonstrates the Qualcomm 2net IoMT Infrastructure components[347]. Components of the Solution:

1. **2net Hub:** A Plug-n-Play device that's installed at a patient's home to interact with existing medical devices and collect data over short and wide range radio

2. **2net Mobile:** A software platform that can be integrated with any third-party mobile application, turning devices into gateways used for capturing and transmitting data for patients at home or on the go

3. **2net Platform:** HIPAA Compliant, FDA-listed, Class, MDDS and CE-marked Class 1 Medical Device (EU) enterprise-grade infrastructure; a cloud-based system that enables E2E medical device data connectivity,

transmission, and integration with any application or portal to serve multiple scenarios.

11.5.11.4 Data Flow

1. Device-specific "agents" are installed on the 2net Hub. These agents initiate data transfers from the devices using short-range radios (e.g., BT, BTLE, Wi-Fi, etc.).

2. Data is uploaded to the 2net Cloud Platform over the cellular network. The data is then transmitted over authenticated SSL connections.

3. For data delivery, the 2net Cloud stores the encrypted data for transmission to the customers.

4. Device data is decrypted and transmitted to the customer through server adapters (including non-standard interfaces) or customer applications.

Products Available for:

1. **MedTech:** Solution for unlocking existing medical device data through 2net framework. Interoperability and connecting Medical Devices is achieved through the 2net Platform.

2. **Hospital/Health System:** Solutions enable scalable chronic care model from hospital to home by providing post-acute care data in a near and real-time environment, helping hospitals to take care of the patients remotely.

3. **Payers:** Solution to allow payers to measure the efficacy of disease management programs and to stratify members for more personalized interventions.

4. **Pharmaceuticals:** The Integration of power drug delivery modules and diagnostic devices through IoT simplifies clinical trials for patients and study investigators, enabling improved adherence and effortless experiences for chronic disease patients, as well as streamlining their experience.

5. **Pharmacy:** 2net Platform helps pharmacists, consumers and their devices to be connected, in order to provide better medical services to the patients. Creating nearer and real-time experience for the patients resulting in personalized experience and allows pharmacist and coaches to tailor therapy and dosage concerning daily activity and monitoring.

11.5.11.5 Azure IoT Suite

The Azure IoT Hub is a messaging system built on top of Azure Event Hub. According to Azure documentation, Azure IoT can receive millions of messages per second that can then be processed at a later time. The Azure IoT hub:

- Enables device-to-cloud, and cloud-to-device communication

- Can store and query device information via "device twins"

- Supports MQTT, MQTT over WebSockets, Advanced Message Queuing Protocol (AMQP), AMQP over WebSockets, and HTTP (also supports custom protocols via Azure IoT protocol gateway)

- Provides per-device identity, and revocable access control

- Is optimized to support millions of simultaneously connected devices

Azure IoT supports device-to-cloud and cloud-to-device communication, including messaging, file transfers, and request-reply methods. For device connectivity challenges, some of the below options could be useful:

- **Device twins:** The Azure IoT hub consists of JSON documents for each device connected to the Azure IoT hub in order to store device state information like metadata, configurations, and conditions. These documents are called "device twins" and can be used to store, synchronize, and query device metadata and state information.

- **Per-device authentication and secure connectivity:** Each device can be provisioned with its security key to enable its connection with the IoT hub. The IoT hub has an identity register to store device identities and keys. Based on this, any backend solution can be configured to maintain a list of devices that should be allowed or rejected. Route device-to-cloud messages to Azure services based on declarative rules: Device-to-cloud message routes can be controlled based on custom rules. A custom post-intake message dispatcher can configure these rules without writing any code.

- **Monitoring of device connectivity operations:** Device identity and connectivity monitoring can be easily achieved via detailed operational logs about identity management and connectivity events. This ability can be used to track issues like too frequent messages, connecting with the wrong credentials, rejecting cloud-to-device messages, etc.

- **An extensive set of device libraries:** Azure IoT provides an SDK for various platforms in various languages, such as C for Linux, Windows, and other real-time operating systems as well as managed languages like C#, Java, and JavaScript for ease of use.

- **IoT protocols and extensibility:** In case your solution cannot use device libraries, the IoT hub supports protocols like MQTT, MQTT over WebSockets, AMQP, AMQP over WebSockets, and HTTP. It also supports custom protocols via the Azure IoT protocol gateway to make connectivity easy.

- **Scale:** The Azure IoT hub can receive millions of messages per second, from millions of connected devices, to produce numerous events.

- **Compliance:** Azure is compliant with the Health Insurance Portability and Accountability Act (HIPPA/HITECH), Health Information Trust Alliance (HITRUST), Common Security Framework (CSF), and Minimum Acceptable Risk Standards for Exchanges (MARS-E).

11.5.11.6 Intel IoT

Intel has a history of providing industrial solutions in healthcare, from medical devices to medical software for disease monitoring and management. Now the company is moving on to next-generation IoMT solutions that can address the requirements of different healthcare areas (e.g., remote monitoring). Intel has developed reference architectures to be used in the healthcare sector to fulfil the needs of devices, data, services, security, and analytics. Because of the company's healthcare industry experience, it already has various toolkits to build IoMT solutions for monitoring and managing healthcare-based services. The Intel IoMT platform can support millions of connected and unconnected medical devices, as well as push data to the Intel cloud for batch and near real-time processing.

The Intel IoT platform provides vendors with various toolkits to build end-to-end, IoT-based solutions. Intel has defined reference architectures for the IoT solution, which addresses requirements for data and device security, device discovery, provisioning and management, data normalization, analytics, and services. These reference architectures are designed for two use cases: (1) connecting legacy infrastructure (i.e., "connecting the unconnected"), and (2) building infrastructure for smart and connected devices, as represented in the below architecture specification. This architecture specification is intended to support developers, vendors, and service providers as they develop and deploy IoMT solutions with five key tenets defined:

1. Discover and provide help to automate device setup from things-to-cloud in minutes to ease deployment

2. Provide services to support data and device management from things-to-cloud

3. Ingest data and control devices to support interoperability, thus facilitating the normalization of protocols and data formats

4. Secure the IoMT platform and platform-based applications at the hardware and software level, including data and devices

5. Provide customer value through insightful, real-time analytics from things-to-cloud

Intel's IoT vision is to sense generated data from many devices, collect and communicate that data, analyze the data and turn it into actionable insight, and use that insight to optimize how things work. Intel has various components to create a robust IoMT solution, and it also can collect data from devices not connected to the Internet. Specifically, there is a developer-enabling layer (i.e., Application programming interface (API), Software development kit (SDK) and Dev Tools) that helps bring those devices within the scope to collect data and share it across devices and services. Intel has defined layers in its reference architecture for management, security, communication, and data. A strategic partnership between Intel and Statistical Analysis System (SAS) ensures the delivery of unprecedented performance and simplicity in order to provide access to powerful SAS Analytics. This partnership has enabled Intel customers to use the SAS analytics capabilities to deliver both real-time and predictive analytics.

11.5.12 Compliance and Regulations

In the healthcare industry, protected health information (PHI) is the primary focus of information technology (IT) risk management practices. Traditionally, sensitive health information has been tracked and managed via paper records. The maintenance and protection of these records relied heavily on people and processes. The risk of unauthorized access, modification, or destruction of information existed, but on average, it would impact just a handful of individuals. The increased adoption of technology in healthcare (e.g., electronic medical records, health information exchanges, networked medical devices) increases the risk of PHI becoming vulnerable to unauthorized disclosures because of the greater amount of data now accessible.

As per BitSight's security rating[121], healthcare is one of the more poorly performing industries. Healthcare data is susceptible and private, so any IoMT solution has the utmost responsibility to secure patient information. However, some challenges for IoMT are bigger than just protecting stored patient information. For example, the transmission of data from device-to-cloud storage through a network and its intermediate points also needs to be secured. Most importantly, healthcare solutions must compy with several compliance standards to ensure the security of PHI:

11.5.12.1 HIPAA Rules

HIPPA was introduced in 1996 and applies to healthcare providers involved in electronic transactions (e.g., health plans, etc.), as well as service providers granting access to any third party associates.

11.5.12.2 HITECH Act

Introduced in 2009, this act extended HIPPA to a set of federal standards intended to protect the security and privacy of PHI. HIPAA and HITECH impose requirements related to the use and disclosure of PHI, appropriate safeguards to protect PHI, individual rights, and administrative responsibilities.

11.5.12.3 HITRUST

Both the Health Information Trust Alliance (HITRUST), and the HITRUST Common Security Framework (CSF) are recent additions to the healthcare information security and compliance landscape. HITRUST was established in 2008 to enable trust in the healthcare industry. The CSF is a framework designed to provide prescriptive, comprehensive guidance for implementing reasonable and appropriate security controls based on risk and agreed to by the broader industry.

11.5.12.4 PCI

The Payment Card Industry (PCI) and the PCI Data Security Standard (DSS) are more broadly focused, international industry groups that have set requirements for payment card (e.g., credit card) processors and the merchants accepting those cards. The reason this compliance is important is that many healthcare organizations, such as hospitals and physician practices, accept credit cards as a form of payment for healthcare services. In a typical healthcare monitoring system, the cloud is the preferred platform to aggregate, store, and analyze data collected from the Medical Internet of Things (or MIoT) devices used by patients or medical facilities. However, remote cloud servers and storage can be a source of delay and due to distant communication and networking. Particularly, in case of an emergency, a minor delay or response on analyzed data can result in inaccurate treatment decisions that may put the patient's life at risk. An intermediate layer of fog or edge nodes are used to overcome network and communication delays and storage of MIoT data. To this end, an association of MIoT devices, fog computing and cloud computing have now become the most preferred solution for a typical healthcare monitoring system.

11.5.13 What We See in Future

The following topics cover the broad area of IoT and smart healthcare future trends.

11.5.13.1 Healthcare Robots

What if the person providing personal care at home is not a person at all but a robot instead? Japan, with its One Child per Family policy, faces a much more significant ageing problem than we do, and they are turning to healthcare robots for help. Japan is pioneering the early development of healthcare robots and in a survey conducted in 2018 by Orix Living Corp., over 80 percent of people said they are ready to or want to receive nursing care from robots. South Korea also faces an ageing issue, as does most of the globe, and they have mandated a robot in every home by 2020. They do this because robots can easily do repetitive tasks or ones that are too dangerous or difficult for humans. They can easily lift and transfer heavy patients, but their strong arms lack sensitivity today. That will be fixed with improved sensors. What's especially attractive about robots is that they can work 24/7 without complaining about low wages or the lack of benefits; and while they're expensive to buy now, those costs will fall as technology allows. Robots can be directed by humans or made to learn and operate on their own. Moreover, they can serve as personal assistants (e.g., Roomba vacuum or Paro companion robot seal), can be something we wear or ride in (e.g., exoskeleton or Google's self-driving car), or even something inside our bodies (nano-scale bots in our bloodstream).

11.5.13.2 The Brain-Computer Interface

The ability to sense electrical activity of nerve endings is already leading to advanced prosthetics for amputees, and so people with quadriplegia can control robot arms just by thinking about it. A brain-computer interface could also be used to control an exoskeleton or robot, and the military is already envisioning soldiers with telepathic helmets by 2020. So what might be the result of converging Information and Cognitive Computing? Futurist Ray Kurzweil has studied this field and foresees a supercomputer exceeding the computational and analytically power of the human brain in the early of 2013 and now in 2020 the research going to next level by leveraging Artificial intelligence and machine learning. In many ways, IBM's Watson already has. However, by simply extrapolating Moore's Law into the future, Kurzweil predicts that by 2023, a $1,000 computer will have the power of the human brain and by 2037 a $0.01 computer will too. By 2049 (still possible in my lifetime), a $1,000 computer will exceed the power of the human race, and ten years later a $0.01 computer will. Way before then, we will see improvements in the brain-computer interface, so changes in healthcare beyond 10-20 years get much harder to imagine. Fig.11.4 represents the human brain-computer interfacing. By the end of 2020s, most diseases will go away as

Source: NYC MEDIA LAB, Human-Computer Interaction
Figure 11.4: The Brain-Computer Interface

nanobots become smarter than current medical technology. Nanosystems can replace normal human eating. The Turing test begins to be passable. Self-driving cars begin to take over the roads, and people will not be allowed to drive on highways. By the 2030s, virtual reality will begin to feel 100% real. We will be able to upload our mind/consciousness by the end of the decade. By the 2040s, non-biological intelligence will be a billion times more capable than biological intelligence. Nanotech froglets will be able to make food out of thin air and create any object in the physical world at a whim. By 2045, we will multiply our intelligence a billionfold by linking wirelessly from our neocortex to a synthetic neocortex in the cloud.

Bibliography

[1] Apple support: About touch id advanced security technology. Technical report.

[2] CareKit.

[3] Continuity of Care Record Storage - HealthVault Development.

[4] *The Creative Destruction of Medicine: How the Digital Revolution Will Create Better Health Care.* New York, United States.

[5] Discover, interact, and optimize for smarter healthcare with bpm powered by smart soa.

[6] Epic Interoperability Creates One Virtual System Worldwide.

[7] Healthcare - Health Records.

[8] Iot healthcare solutions - medical internet of things for healthcare devices and smart hospitals.

[9] Merlin.net™ Patient Care Network | Abbott Cardiovascular.

[10] Microsoft HealthVault.

[11] Office-based Physician Electronic Health Record Adoption.

[12] Research.

[13] The value of analytics in healthcare.

[14] watchOS 6 advances health and fitness capabilities for Apple Watch.

[15] What is an electronic health record (EHR)? | HealthIT.gov.

[16] Google Health Data API CCR Reference - Google Health Data API - Google Code, February 2012.

[17] Cisco global cloud index: Forecast and methodology, white paper, Nov 2018.

[18] M. Aazam and E. Huh. Fog computing and smart gateway based communication for cloud of things. In *2014 International Conference on Future Internet of Things and Cloud*, pages 464–470, 2014.

[19] Hezam Akram Abdul-Ghani and Dimitri Konstantas. A comprehensive study of security and privacy guidelines, threats, and countermeasures: An iot perspective. *Journal of Sensor and Actuator Network*, 8(22).

[20] Hezam Akram Abdul-Ghani, Dimitri Konstantas, and Mohammed Mahyoub. A comprehensive iot attacks survey based on a building-blocked reference model. *International Journal of Advanced Computer Science and Applications*, 9(3), 2018.

[21] Mohd Azlan Abu, Siti Fatimah Nordin, Mohd Zubir Suboh, Mohd Syazwan Md Yid, and Aizat Faiz Ramli. Design and development of home security systems based on internet of things via favoriot platform. *International Journal of Applied Engineering Research*, 13(2):1253–1260, 2018.

[22] S. K. Addya, A. Satpathy, B. C. Ghosh, S. Chakraborty, and S. K. Ghosh. Power and time aware vm migration for multi-tier applications over geo-distributed clouds. In *2019 IEEE 12th International Conference on Cloud Computing (CLOUD)*, pages 339–343, 2019.

[23] Swati Agarwal, Shashank Yadav, and Arun Kumar Yadav. An efficient architecture and algorithm for resource provisioning in fog computing. 2016.

[24] B. Ahlgren, C. Dannewitz, C. Imbrenda, D. Kutscher, and B. Ohlman. A survey of information-centric networking. *IEEE Communications Magazine*, 50(7):26–36, July 2012.

[25] Muhammad Aurangzeb Ahmad, Carly Eckert, and Ankur Teredesai. Interpretable machine learning in healthcare. In *Proceedings of the 2018 ACM International Conference on Bioinformatics, Computational Biology, and Health Informatics*, pages 559–560. ACM, 2018.

[26] E. S. A. Ahmed and R. A. Saeed. A Survey of Big Data Cloud Computing Security. *International Journal of Computer Science and Software Engineering (IJCSSE)*, 3:78–85.

[27] Monjur Ahmed. *Ki-Nga-Kopuku: A Decentralised, Distributed Security Model for Cloud Computing*. PhD thesis, Auckland University of Technology, 2018.

[28] Monjur Ahmed, Alan T. Litchfield, and Shakil Ahmed. A generalized threat taxonomy for cloud computing. ACIS, 2014.

[29] Nikolay Akatyev and Joshua I. James. Evidence identification in iot networks based on threat assessment. *Future Generation Computer Systems*, 2017.

[30] A. Al-Fuqaha, M. Guizani, M. Mohammadi, M. Aledhari, and M. Ayyash. Internet of things: A survey on enabling technologies, protocols, and

applications. *IEEE Communications Surveys Tutorials*, 17(4):2347–2376, Fourthquarter 2015.

[31] Ala Al-Fuqaha, Mohsen Guizani, Mehdi Mohammadi, Mohammed Aledhari, and Moussa Ayyash. Internet of things: A survey on enabling technologies, protocols, and applications. *IEEE Communications Surveys & Tutorials*, 17(4):2347–2376, 2015.

[32] Y. Al-Nidawi, H. Yahya, and A. H. Kemp. Impact of mobility on the iot mac infrastructure: Ieee 802.15.4e tsch and lldn platform. In *2015 IEEE 2nd World Forum on Internet of Things (WF-IoT)*, pages 478–483, Dec 2015.

[33] Aftab Ali and Farrukh Aslam Khan. Key agreement schemes in wireless body area networks: Taxonomy and state-of-the-art. *Journal of Medical Systems*, 39(10):115, 2015.

[34] Bako Ali and Ali Awad. Cyber and physical security vulnerability assessment for iot-based smart homes. *Sensors*, 18(3):817, 2018.

[35] Sabir S. Ali, I. and Z. Ullah. Internet of things Security, Device Athentication, and Access Control: A Review. *Int. J. of Computer Science and Information Security (IJCSIS)*, 14(8):456–466, 2016.

[36] H.A. Ali, Z.H. Ali and M.M. Badawy. Internet of Things (IoT): Definitions, Challenges and Recent Research Direction. *International Journal of Computer Applications*, 128(1), 2015.

[37] Fritz Allhoff and Adam Henschke. The internet of things: Foundational ethical issues. *Internet of Things*, 1:55–66, 2018.

[38] Israa Alqassem and Davor Svetinovic. A taxonomy of security and privacy requirements for the internet of things (iot). In *2014 IEEE International Conference on Industrial Engineering and Engineering Management*. IEEE.

[39] Ebraheim Alsaadi and Abdallah Tubaishat. Internet of things: features, challenges, and vulnerabilities. *International Journal of Advanced Computer Science and Information Technology*, 4(1):1–13, 2015.

[40] M. Amadeo, C. Campolo, A. Iera, and A. Molinaro. Named data networking for iot: An architectural perspective. In *2014 European Conference on Networks and Communications (EuCNC)*, pages 1–5, June 2014.

[41] M. Amadeo, C. Campolo, A. Iera, and A. Molinaro. Information centric networking in iot scenarios: The case of a smart home. In *2015 IEEE International Conference on Communications (ICC)*, pages 648–653, June 2015.

[42] M. Amadeo, C. Campolo, and A. Molinaro. Internet of things via named data networking: The support of push traffic. In *2014 International Conference and Workshop on the Network of the Future (NOF)*, pages 1–5, Dec 2014.

[43] Muhammad N. Aman, Kee Chaing Chua, and Biplab Sikdar. Position paper: Physical unclonable functions for iot security. In *Proceedings of the 2nd ACM international workshop on IoT privacy, trust, and security*, pages 10–13. ACM, 2016.

[44] Mahmoud Ammar, Giovanni Russello, and Bruno Crispo. Internet of things: A survey on the security of iot frameworks. *Journal of Information Security and Applications*, 38:8–27, 2018.

[45] A. Amokrane, M. F. Zhani, R. Langar, R. Boutaba, and G. Pujolle. Greenhead: Virtual data center embedding across distributed infrastructures. *IEEE Transactions on Cloud Computing*, 1(1):36–49, 2013.

[46] Theodoros Anagnostopoulos, Arkady Zaslavsy, Alexey Medvedev, and Sergei Khoruzhnicov. Top–k query based dynamic scheduling for iot-enabled smart city waste collection. In *2015 16th IEEE International Conference on Mobile Data Management*, volume 2, pages 50–55. IEEE, 2015.

[47] Ioannis Andrea, Chrysostomos Chrysostomou, and George Hadjichristofi. Internet of things: Security vulnerabilities and challenges. In *2015 IEEE Symposium on Computers and Communication (ISCC)*, pages 180–187. IEEE, 2015.

[48] Kishore Angrishi. Turning internet of things (iot) into internet of vulnerabilities (iov): Iot botnets. *arXiv preprint arXiv:1702.03681*, 2017.

[49] I. Ari, B. Hong, E. L. Miller, S. A. Brandt, and D. D. E. Long. Managing flash crowds on the internet. In *11th IEEE/ACM International Symposium on Modeling, Analysis and Simulation of Computer Telecommunications Systems, 2003. MASCOTS 2003*, pages 246–249, Oct 2003.

[50] S. Arshad, M. A. Azam, M. H. Rehmani, and J. Loo. Recent advances in information-centric networking-based internet of things (icn-iot). *IEEE Internet of Things Journal*, 6(2):2128–2158, April 2019.

[51] M. D. Asuncao, R. N. Calheiros, S. Bianchi, M. A. S. Netto, and R. Buyya. Big Data Computing and Clouds: Trends and future directions. *J. Parallel Distributed Computing*, 79:3–15, 2015.

[52] Cindy Atoji. Markle Director Says PHR Fragmentation Is a Problem. page 3.

[53] Luigi Atzori, Antonio Iera, and Giacomo Morabito. The internet of things: A survey. *Computer Networks*, 54(15):2787–2805, 2010.

[54] Luigi Atzori, Antonio Iera, and Giacomo Morabito. The internet of things: A survey. *Computer networks*, 54(15):2787–2805, 2010.

[55] O. Avodele, A. A. Izang, S. O. Kuyoro, and F. Y. Osisanwo. Big Data and Cloud Computing Issues. *Int. J. of Computer Applications*, 133(12):14–19, 2015.

[56] Sachin Babar, Antonietta Stango, Neeli Prasad, Jaydip Sen, and Ramjee Prasad. Proposed embedded security framework for internet of things (iot). In *2011 2nd International Conference on Wireless Communication, Vehicular Technology, Information Theory and Aerospace & Electronic Systems Technology (Wireless VITAE)*, pages 1–5. IEEE, 2011.

[57] Emmanuel Baccelli, Christian Mehlis, Oliver Hahm, Thomas C. Schmidt, and Matthias Wahlisch. Information centric networking in the iot: Experiments with ndn in the wild. In *Proceedings of the 1st ACM Conference on Information-Centric Networking*, ACM-ICN '14, pages 77–86, New York, NY, USA, 2014. ACM.

[58] Debasis Bandyopadhyay and Jaydip Sen. Internet of things: Applications and challenges in technology and standardization. *Wireless Personal Communications*, 58(1):49–69, 2011.

[59] Sujogya Banerjee, Shahrzad Shirazipourazad, and Arunabha Sen. On region-based fault tolerant design of distributed file storage in networks. In *2012 Proceedings IEEE INFOCOM*, pages 2806–2810. IEEE, 2012.

[60] M.B. Barcena and C. Wueest. Insecurity in the internet of things - symantec. 2015.

[61] Jane H. Barnsteiner. Medication Reconciliation: Transfer of Medication Information Across Settings - Keeping It Free from Error. *AJN The American Journal of Nursing*, 105(3):31, March 2005.

[62] D. Bender and K. Sartipi. Hl7 FHIR: An Agile and RESTful Approach to Healthcare Information Exchange. In *Proceedings of the 26th IEEE International Symposium on Computer-Based Medical Systems*, pages 326–331, June 2013.

[63] Andrea Beratarrechea, Allison G. Lee, Jonathan M. Willner, Eiman Jahangir, Agustín Ciapponi, and Adolfo Rubinstein. The Impact of Mobile Health Interventions on Chronic Disease Outcomes in Developing Countries: A Systematic Review. *Telemedicine and e-Health*, 20(1):75–82, November 2013.

[64] Abdur Rahim Biswas and Raffaele Giaffreda. Iot and cloud convergence: Opportunities and challenges. In *2014 IEEE World Forum on Internet of Things (WF-IoT)*, pages 375–376. IEEE, 2014.

[65] L. F. Bittencourt, M. M. Lopes, I. Petri, and O. F. Rana. Towards virtual machine migration in fog computing. In *2015 10th International Conference on P2P, Parallel, Grid, Cloud and Internet Computing (3PGCIC)*, pages 1–8, 2015.

[66] David Blumenthal. Launching HITECH. *New England Journal of Medicine*, 362(5):382–385, February 2010.

[67] B. Ohlman, D. Oran, and D. Kutscher. Information-centric networking (icnrg),ietf.

[68] Flavio Bonomi, Rodolfo Milito, Jiang Zhu, and Sateesh Addepalli. Fog computing and its role in the internet of things. In *Proceedings of the First Edition of the MCC Workshop on Mobile Cloud Computing*, MCC '12, 2012.

[69] Eleonora Borgia. The internet of things vision: Key features, applications and open issues. *Computer Communications*, 54:1–31, 2014.

[70] Taxiarchis Botsis, Gunnar Hartvigsen, Fei Chen, and Chunhua Weng. Secondary Use of EHR: Data Quality Issues and Informatics Opportunities. *Summit on Translational Bioinformatics*, 2010:1–5, March 2010.

[71] Alessio Botta, Walter De Donato, Valerio Persico, and Antonio Pescapé. Integration of cloud computing and internet of things: a survey. *Future generation Computer Systems*, 56:684–700, 2016.

[72] Maged N. Kamel Boulos, Steve Wheeler, Carlos Tavares, and Ray B. Jones. How smartphones are changing the face of mobile and participatory healthcare: an overview, with example from ecaalyx. In *Biomedical Engineering Online*, 2011.

[73] Rihab Boussada, Mohamed Elhoucine Elhdhili, and Leila Azouz Saidane. Privacy preserving solution for internet of things with application to ehealth. In *2017 IEEE/ACS 14th International Conference on Computer Systems and Applications (AICCSA)*, pages 384–391. IEEE, 2017.

[74] Stephen W Boyd and Angelos D Keromytis. Sqlrand: Preventing sql injection attacks. In *International Conference on Applied Cryptography and Network Security*, pages 292–302. Springer, 2004.

[75] Tim Brennan and William L Oliver. Emergence of machine learning techniques in criminology: Implications of complexity in our data and in research questions. *Criminology & Pub. Pol'y*, 12:551, 2013.

[76] Gerben Broenink, Jaap-Henk Hoepman, Christian van 't Hof, Rob van Kranenburg, David Smits, and Tijmen Wisman. The Privacy Coach: Supporting customer privacy in the Internet of Things. *arXiv:1001.4459 [cs]*, Jan 2010. arXiv: 1001.4459.

[77] R. Brzoza-Woch, M. Konieczny, P. Nawrocki, T. Szydlo, and K. Zielinski. Embedded systems in the application of fog computing - levee monitoring use case. In *2016 11th IEEE Symposium on Industrial Embedded Systems (SIES)*, 2016.

[78] Jenna Burrell. How the machine 'thinks': Understanding opacity in learning algorithms machine learning algorithms. *Big Data & Society*, 3(1):2053951 715622512.

[79] Fan Cai, Nhien-An Le-Khac, and Tahar Kechadi. Clustering approaches for financial data analysis: A survey. *arXiv preprint arXiv:1609.08520*, 2016.

[80] V. Cardellini, V. Grassi, F. L. Presti, and M. Nardelli. On qos-aware scheduling of data stream applications over fog computing infrastructures. In *2015 IEEE Symposium on Computers and Communication (ISCC)*, pages 271–276, 2015.

[81] Luciana Cardoso, Fernando Marins, Filipe Portela, Manuel Santos, António Abelha, and José Machado. The Next Generation of Interoperability Agents in Healthcare. *International Journal of Environmental Research and Public Health*, 11(5):5349–5371, May 2014.

[82] Percivil M Carrera and John F.P. Bridges. Globalization and healthcare: Understanding health and medical tourism. *Expert Review of Pharmacoeconomics & Outcomes Research*, 6(4):447–454, 2006. PMID: 20528514.

[83] Emiliano Casalicchio and Luca Silvestri. Mechanisms for sla provisioning in cloud-based service providers. *Computer Networks*, 57:795–810, 2013.

[84] Valentina Casola, Antonio Cuomo, Massimiliano Rak, and Umberto Villano. The cloudgrid approach: Security analysis and performance evaluation. *Future Generation Computer Systems*, 29:387–401, 2013.

[85] V. Cerf. Access Control and the Internet of Things. *Int. J. of Computer science and Information Security (IJCSIS)*, 19(5):96, 2015.

[86] S. Challa, M. Wazid, A. K. Das, and M. K. Khan. Authentication Protocols for Implantable Medical Devices: Taxonomy, Analysis and Future Directions. *IEEE Consumer Electronics Magazine*, 7(1):57–65, January 2018.

[87] Moumena A Chaqfeh and Nader Mohamed. Challenges in middleware solutions for the internet of things. In *2012 International Conference on Collaboration Technologies and Systems (CTS)*, pages 21–26. IEEE, 2012.

[88] Sanjit Chatterjee and Palash Sarkar. *Identity-based encryption*. Springer Science & Business Media, 2011.

[89] Dong Chen, Guiran Chang, Dawei Sun, Jiajia Li, Jie Jia, and Xingwei Wang. Trm-iot: A trust management model based on fuzzy reputation for internet of things. *Computer Sci. Inf. Syst.*, 8(4):1207–1228, 2011.

[90] Kejun Chen, Shuai Zhang, Zhikun Li, Yi Zhang, Qingxu Deng, Sandip Ray, and Yier Jin. Internet-of-things security and vulnerabilities: Taxonomy, challenges, and practice. *Journal of Hardware and Systems Security*, 2:97–110.

[91] Le Chen, Zhenfu Cao, Rongxing Lu, Xiaohui Liang, and Xuemin Shen. EPF: An event-aided packet forwarding protocol for privacy-preserving mobile healthcare social networks. In *2011 IEEE Global Telecommunications Conference-GLOBECOM 2011*, pages 1–5. IEEE, 2011.

[92] Min Chen, Sergio Gonzalez, Athanasios Vasilakos, Huasong Cao, and Victor C. Leung. Body area networks: A survey. *Mob. Netw. Appl.*, 16(2):171–193, April 2011.

[93] Yanpei Chen, Vern Paxson, and Randy H Katz. What's new about cloud computing security. *University of California, Berkeley Report No. UCB/EECS-2010-5 January*, 20(2010):2010–5, 2010.

[94] Jun Cheng, Wei Huang, Shuangliang Cao, Ru Yang, Wei Yang, Zhaoqiang Yun, Zhijian Wang, and Qianjin Feng. Enhanced performance of brain tumor classification via tumor region augmentation and partition. *PLOS ONE*, 2015.

[95] M. Chiang and T. Zhang. Fog and iot: An overview of research opportunities. *IEEE Internet of Things Journal*, 3(6):854–864, Dec 2016.

[96] Shin-Yan Chiou and Zhen-Yuan Liao. A real-time, automated and privacy-preserving mobile emergency-medical-service network for informing the closest rescuer to rapidly support mobile-emergency-call victims. *IEEE Access*, 6:35787–35800, 2018.

[97] Richard Chow, Markus Jakobsson, Ryusuke Masuoka, Jesus Molina, Yuan Niu, Elaine Shi, and Zhexuan Song. Authentication in the clouds: A framework and its application to mobile users. In *ACM CCSW'10*, pages 1–6, Chicago, Illinois, USA, October 2010. ACM.

[98] N. M. M. K. Chowdhury, M. R. Rahman, and R. Boutaba. Virtual network embedding with coordinated node and link mapping. In *IEEE INFOCOM 2009*, 2009.

[99] Francis Collins. The Real Promise of Mobile Health Apps.

[100] Diane Cook, Glen Duncan, Gina Sprint, and Roschelle Fritz. Using smart city technology to make healthcare smarter. *Proceedings of the IEEE*, pages 1–15, 01 2018.

[101] Mathieu Cunche, Daniel Le Métayer, and Victor Morel. A Generic Information and Consent Framework for the IoT (Extended Version), page 21.

[102] B. Dab, I. Fajjari, N. Aitsaadi, and G. Pujolle. Vnr-ga: Elastic virtual network reconfiguration algorithm based on genetic metaheuristic. In *2013 IEEE Global Communications Conference (GLOBECOM)*, 2013.

[103] Christian Dameff, Brian Clay, and Christopher A. Longhurst. Personal Health Records: More Promising in the Smartphone Era? *JAMA*, 321(4):339–340, January 2019.

[104] Amir Vahid Dastjerdi and Rajkumar Buyya. Fog computing: Helping the internet of things realize its potential. *Computer*, 49(8):112–116, 2016.

[105] S. K. Datta, C. Bonnet, and J. Haerri. Fog computing architecture to enable consumer centric internet of things services. In *2015 International Symposium on Consumer Electronics (ISCE)*, 2015.

[106] David W. Dawn, X.S. and P. Adrian. In *IEEE Symposium on Security and Privacy*, pages 44–55.

[107] Ahmed Dawoud, Seyed Shahristani, and Chun Raun. Deep learning and software-defined networks: Towards secure iot architecture. *Internet of Things*, 3:82–89, 2018.

[108] G.M. de Brito, P.B. Velloso, and I.M. Moraes. *Information-Centric Networks: A New Paradigm for the Internet*. Wiley, 2013.

[109] Jyoti Deogirikar and Amarsinh Vidhate. Security attacks in iot: A survey. In *2017 International Conference on I-SMAC (IoT in Social, Mobile, Analytics and Cloud)(I-SMAC)*, pages 32–37. IEEE, 2017.

[110] Catherine M. DesRoches, Eric G. Campbell, Sowmya R. Rao, Karen Donelan, Timothy G. Ferris, Ashish Jha, Rainu Kaushal, Douglas E. Levy, Sara Rosenbaum, Alexandra E. Shields, and David Blumenthal. Electronic Health Records in Ambulatory Care - A National Survey of Physicians. *New England Journal of Medicine*, 359(1):50–60, July 2008.

[111] Don E. Detmer. Building the national health information infrastructure for personal health, health care services, public health, and research. *BMC Medical Informatics and Decision Making*, 3:1 – 1, 2003.

[112] Andrea Detti, Nicola Blefari Melazzi, Stefano Salsano, and Matteo Pomposini. CONET: A content centric inter-networking architecture. In *Proceedings of the ACM SIGCOMM Workshop on Information-centric Networking*, ICN '11, pages 50–55, New York, NY, USA, 2011. ACM.

[113] B. Di Martino, M. Rak, M. Ficco, A. Esposito, S.A. Maisto, and S. Nacchia. Internet of things reference architectures, security and interoperability: A survey. *Internet of Things*, 1:99–112, 2018.

[114] D. Dietrich, A. Rizk, and P. Papadimitriou. Multi-provider virtual network embedding with limited information disclosure. *IEEE Transactions on Network and Service Management*, 12(2):188–201, 2015.

[115] Vladimir Dimitrov and Ventzislav Koptchev. Psirp project – publish-subscribe internet routing paradigm: New ideas for future internet. In *Proceedings of the 11th International Conference on Computer Systems and Technologies and Workshop for PhD Students in Computing on International Conference on Computer Systems and Technologies*, CompSysTech '10, pages 167–171, New York, NY, USA, 2010. ACM.

[116] I. U. Din, S. Hassan, M. K. Khan, M. Guizani, O. Ghazali, and A. Habbal. Caching in information-centric networking: Strategies, challenges, and future research directions. *IEEE Communications Surveys Tutorials*, 20(2):1443–1474, Secondquarter 2018.

[117] Maman Dogba, Anara Dossa, Erik Breton, and Ruth Gandonou-Migan. Using information and communication technologies to involve patients and the public in health education in rural and remote areas: A scoping review. *BMC Health Services Research*, 19, 12, 2019.

[118] Jiang Dong, Dafang Zhuang, Yaohuan Huang, and Jingying Fu. Advances in multi-sensor data fusion: Algorithms and applications. *Sensors*, 9(10):7771–7784, 2009.

[119] Penglin Dong, Weichao Wang, Xinghua Shi, and Tuanfa Qin. Lightweight key management for group communication in body area networks through physical unclonable functions. In *Proceedings of the Second IEEE/ACM International Conference on Connected Health: Applications, Systems and Engineering Technologies*, pages 102–107. IEEE Press, 2017.

[120] L. Donnell. Iot security concerns peaking with no end in sight. In *HackerOne CEO Talks Bug Bounty Programs at RSA Conference*, 2018.

[121] Susan Doyle-Lindrud. The evolution of the electronic health record. *Clinical Journal of Oncology Nursing*, 19:153–4, 04, 2015.

[122] Adrian Duncan, Sadie Creese, and Michael Goldsmith. Insider attacks in cloud computing. In *IEEE 11th International Conference on Trust, Security and Privacy in Computing and Communications*, pages 857–862. IEEE, 2012.

[123] Otmane El Mouaatamid, Mohammed Lahmer, and Mostafa Belkasmi. Internet of things security: Layered classification of attacks and possible countermeasures. *Electronic Journal of Information Technology*, (9), 2016.

[124] Nadia El Mrabet and Marc Joye. *Guide to pairing-based cryptography*. Chapman and Hall/CRC, 2017.

[125] D Evans. Iot by cisco 2011. *Cisco Internet Bus. Solut. Gr.*, April, 2011.

[126] I. Fajjari, N. Aitsaadi, G. Pujolle, and H. Zimmermann. Vne-ac: Virtual network embedding algorithm based on ant colony metaheuristic. In

2011 IEEE International Conference on Communications (ICC), pages 1–6, 2011.

[127] Chun-I Fan and Shi-Yuan Huang. Controllable privacy preserving search based on symmetric predicate encryption in cloud storage. *Future Gener Comput Syst*, 29:1716–1724, 2013.

[128] Bahar Farahani, Farshad Firouzi, Victor I. Chang, Mustafa Badaroglu, Nicholas Constant, and Kunal Mankodiya. Towards fog-driven iot ehealth: Promises and challenges of iot in medicine and healthcare. *Future Generation Comp. Syst.*, 78:659–676, 2018.

[129] Diogo A. B. Fernandes, Liliana F. B. Soares, Joao V. Gomes, Mario M. Freire, and Pedro R. M. Inacio. Security issues in cloud environments: A survey. *International Journal of Information Security*, 13:113–170, 2014.

[130] Tiago M. Fernández-Caramés and Paula Fraga-Lamas. Towards the internet of smart clothing: A review on iot wearables and garments for creating intelligent connected e-textiles. *Electronics*, 7(12):405, 2018.

[131] Niroshinie Fernando, Seng W. Loke, and Wenny Rahayu. Mobile cloud computing: A survey. *Future Generation Computer Systems*, 29:84–106, 2013.

[132] Jeffrey M. Ferranti, R. Clayton Musser, Kensaku Kawamoto, and W. Ed Hammond. The Clinical Document Architecture and the Continuity of Care Record: A Critical Analysis. *Journal of the American Medical Informatics Association*, 13(3):245–252, May 2006.

[133] Giancarlo Fortino, Giuseppe Di Fatta, Mukaddim Pathan, and Athanasios V. Vasilakos. Cloud-assisted body area networks: State-of-the-art and future challenges. *Wireless Networks*, 20(7):1925–1938, 2014.

[134] Giancarlo Fortino and Paolo Trunfio. *Internet of things based on smart objects: Technology, middleware and applications.* Springer, 2014.

[135] Ian Foster, Yong Zhao, Ioan Raicu, and Shiyong Lu. Cloud computing and grid computing 360-degree compared. *arXiv preprint arXiv:0901.0131*, 2008.

[136] N. Fotiou and G. C. Polyzos. Realizing the internet of things using information-centric networking. In *10th International Conference on Heterogeneous Networking for Quality, Reliability, Security and Robustness*, pages 193–194, Aug 2014.

[137] Octavian Fratu, Cătălina Monica C. Pena, Razvan Craciunescu, and Simona Halunga. Fog computing system for monitoring mild dementia and copd patients - romanian case study. *2015 12th International Conference on Telecommunication in Modern Satellite, Cable and Broadcasting Services (TELSIKS)*, pages 123–128, 2015.

[138] C. Free, R. Whittaker, R. Knight, T. Abramsky, A. Rodgers, and I. G. Roberts. Txt2stop: A Pilot Randomised Controlled Trial of Mobile Phone-Based Smoking Cessation Support. *Tobacco Control*, 18(2):88–91, April 2009.

[139] Ivan Ganchev, Zhanlin Ji, and Máirtín O'Droma. A generic iot architecture for smart cities. 2014.

[140] A. Gholami and E. Laure. Big Data Security and Privacy Issues in the Cloud. *Int. J. of Network Security and its Applications (IJNSA)*, 8(1):59–79, 2016.

[141] T. N. Gia, M. Jiang, A. Rahmani, T. Westerlund, P. Liljeberg, and H. Tenhunen. Fog computing in healthcare internet of things: A case study on ecg feature extraction. In *2015 IEEE International Conference on Computer and Information Technology; Ubiquitous Computing and Communications; Dependable, Autonomic and Secure Computing; Pervasive Intelligence and Computing*, pages 356–363, 2015.

[142] Stephen Goldsmith and Susan Crawford. *The responsive city: Engaging communities through data-smart governance*. John Wiley & Sons, 2014.

[143] Nelson Gonzalez, Charles Miers, Fernando Redigolo, Marcos Simplicio, Tereza Carvalho, Mats Naslund, and Makan Pourzandi. A quantitative analysis of current security concerns and solutions for cloud computing. *Journal of Cloud Computing*, 1(11):1–18, 2012.

[144] Panagiotis I. Radoglou Grammatikis, Panagiotis G. Sarigiannidis, and Ioannis D. Moscholios. Securing the internet of things: Challenges, threats and solutions. *Internet of Things*, 2018.

[145] Bernd Grobauer, Tobias Walloschek, and Elmar Stocker. Understanding cloud computing vulnerabilities. *IEEE Security & Privacy*, 9(2):50–57, 2010.

[146] Brend Grobauer, Tobias Walloschek, and Elamr Stocker. Understanding cloud computing vulnerabilities. *Cloud Computing*, March/April 2011:50–57, 2011.

[147] X. Guan, B. Choi, and S. Song. Topology and migration-aware energy efficient virtual network embedding for green data centers. In *2014 23rd International Conference on Computer Communication and Networks (ICCCN)*, 2014.

[148] X. Guan, X. Wan, B. Choi, and S. Song. Ant colony optimization based energy efficient virtual network embedding. In *2015 IEEE 4th International Conference on Cloud Networking (CloudNet)*, 2015.

[149] Jayavardhana Gubbi, Rajkumar Buyya, Slaven Marusic, and Marimuthu Palaniswami. Internet of things (iot): A vision, architectural elements, and future directions. *Future Generation Computer Systems*, 29(7):1645–1660, 2013.

[150] D. Gunatilaka. Recent information-centric networking approaches. 2013.

[151] Linke Guo, Chi Zhang, Jinyuan Sun, and Yuguang Fang. A privacy-preserving attribute-based authentication system for mobile health networks. *IEEE Transactions on Mobile Computing*, 13(9):1927–1941, 2013.

[152] Mostafa Haghi, Kerstin Thurow, and Regina Stoll. Wearable Devices in Medical Internet of Things: Scientific Research and Commercially Available Devices. *Healthcare Informatics Research*, 23(1):4–15, January 2017.

[153] Sofiane Hamrioui, Camil Adam Mohamed Hamrioui, Jaime Lioret, and Pascal Lorenz. Smart and self-organised routing algorithm for efficient iot communications in smart cities. *IET Wireless Sensor Systems*, 8(6):305–312, 2018.

[154] Gerhard Hancke et al. Eavesdropping attacks on high-frequency rfid tokens. In *4th Workshop on RFID Security (RFIDSec)*, volume 9440, pages 259–288, 2008.

[155] Mark A. Hanson, Harry C. Powell Jr., Adam T. Barth, Kyle Ringgenberg, Benton H. Calhoun, James H. Aylor, and John Lach. Body area sensor networks: Challenges and opportunities. *Computer*, 42(1):58–65, 2009.

[156] Veli-Pekka Harjola, John Parissis, Hans-Peter Brunner-La Rocca, Jelena Čelutkienė, Ovidiu Chioncel, Sean P. Collins, Daniel De Backer, Gerasimos S. Filippatos, Etienne Gayat, Loreena Hill, Mitja Lainscak, Johan Lassus, Josep Masip, Alexandre Mebazaa, Òscar Miró, Andrea Mortara, Christian Mueller, Wilfried Mullens, Markku S. Nieminen, Alain Rudiger, Frank Ruschitzka, Petar M. Seferovic, Alessandro Sionis, Antoine Vieillard-Baron, Jean Marc Weinstein, Rudolf A. de Boer, Maria G. Crespo-Leiro, Massimo Piepoli, and Jillian P. Riley. Comprehensive in-hospital monitoring in acute heart failure: Applications for clinical practice and future directions for research. A statement from the acute heart failure committee of the heart failure association (hfa) of the european society of cardiology (esc). *European Journal of Heart Failure*, 20(7):1081–1099, 2018.

[157] Ibrahim Abaker Targio Hashem, Ibrar Yaqoob, Nor Badrul Anuar, Salimah Mokhtar, Abdullah Gani, and Samee Ullah Khan. The rise of big data on cloud computing: Review and open research issues. *Information Systems*, 47:98–115, 2015.

[158] Wan Haslina Hassan et al. Current research on internet of things (iot) security: A survey. *Computer Networks*, 148:283–294, 2019.

[159] Vikas Hassija, Vinay Chamola, Vikas Saxena, Divyansh Jain, Pranav Goyal, and Biplab Sikdar. A survey on iot security: Application areas, security threats, and solution architectures. *IEEE Access*, 7:82721–82743, 2019.

[160] D. He and S. Zeadally. An Analysis of RFID Authentication Schemes for Internet of Things in Healthcare Environment Using Elliptic Curve Cryptography. *IEEE Internet of Things Journal*, 2(1):72–83, February 2015.

[161] Debiao He, Sherali Zeadally, Neeraj Kumar, and Jong-Hyouk Lee. Anonymous authentication for wireless body area networks with provable security. *IEEE Systems Journal*, 11(4):2590–2601, 2016.

[162] Juan Hernández-Serrano, Jose L. Muñoz, Olga León, Lars Mikkelsen, Hans-Peter Schwefel, and Arne Broring. Privacy risk analysis in the iot domain. In *2018 Global Internet of Things Summit (GIoTS)*, pages 1–6. IEEE, 2018.

[163] Jason M. Hirst, Jonathan R. Miller, Brent A. Kaplan, and Derek D. Reed. Watts up? pro ac power meter for automated energy recording, 2013.

[164] Kirak Hong, David Lillethun, Umakishore Ramachandran, Beate Ottenwalder, and Boris Koldehofe. Mobile fog: A programming model for large-scale applications on the internet of things. In *Proceedings of the second ACM SIGCOMM Workshop on Mobile Cloud Computing*, pages 15–20. ACM, 2013.

[165] Hong-Kun Zheng, J. Li, Y. Gong, W. Chen, Zhiwen Yu, Z. Zhan, and Ying Lin. Link mapping-oriented ant colony system for virtual network embedding. In *2017 IEEE Congress on Evolutionary Computation (CEC)*, 2017.

[166] M. S. Hossain, G. Muhammad, S. M. M. Rahman, W. Abdul, A. Alelaiwi, and A. Alamri. Toward End-to-End Biomet Rics-Based Security for Iot Infrastructure. *IEEE Wireless Communications*, 23(5):44–51, October 2016.

[167] Md Ekramul Hossain, Arif Khan, Mohammad Ali Moni, and Shahadat Uddin. Use of electronic health data for disease prediction: A comprehensive literature review. *IEEE/ACM Transactions on Computational Biology and Bioinformatics*, pages 1–1, 2019.

[168] X. Hou, Y. Li, M. Chen, D. Wu, D. Jin, and S. Chen. Vehicular fog computing: A viewpoint of vehicles as the infrastructures. *IEEE Transactions on Vehicular Technology*, 65(6), 2016.

[169] Zhongsheng Hou, Jian-Xin Xu, and Jingwen Yan. An iterative learning approach for density control of freeway traffic flow via ramp metering. *Transportation Research Part C: Emerging Technologies*, 16(1):71–97, 2008.

[170] Y. Hsieh, H. Hong, P. Tsai, Y. Wang, Q. Zhu, M. Y. S. Uddin, N. Venkatasubramanian, and C. Hsu. Managed edge computing on internet-of-things devices for smart city applications. In *NOMS 2018 - 2018 IEEE/IFIP Network Operations and Management Symposium*, pages 1–2, 2018.

[171] Ling Huang, Anthony D. Joseph, Blaine Nelson, Benjamin I.P. Rubinstein, and J. D. Tygar. Adversarial Machine Learning. In *Proceedings of the 4th ACM Workshop on Security and Artificial Intelligence*, AISec '11, pages 43–58, New York, NY, USA, 2011. ACM. event-place: Chicago, Illinois, USA.

[172] Qinlong Huang, Licheng Wang, and Yixian Yang. Secure and privacy-preserving data sharing and collaboration in mobile healthcare social networks of smart cities. *Security and Communication Networks*, 2017, 2017.

[173] Jane Hunter. MetaNet - A Metadata Term Thesaurus to Enable Semantic Interoperability Between Metadata Domains. *Journal of Digital Information*, 1(8), January 2006.

[174] Bertin E. Hussein, D. and V. Frey. In 20th *Conference on Innovations in Clouds, Internet and Networks (ICIN)*, pages 7–9.

[175] Jung-Hwan Hwang, Tae-Wook Kang, Youn-Tae Kim, and Seong-Ook Park. Analysis on co-channel interference of human body communication supporting ieee 802.15.6 ban standard. *ETRI Journal*, 37(3):439–449, 2015.

[176] S Intille. The Precision Medicine Initiative and Pervasive Health Research. *IEEE Pervasive Computing*, 15(1):88–91, 2016.

[177] Arsi S. Inukollu, V.N. and S.R. Ravuri. Security Issues Associated with Big Data in Cloud Computing. *Int. J. of Network Security and its Applications (IJNSA)*, 6(3):45–56, 2014.

[178] M. M. Islam, M. A. Razzaque, M. M. Hassan, W. N. Ismail, and B. Song. Mobile cloud-based big healthcare data processing in smart cities. *IEEE Access*, 5:11887–11899, 2017.

[179] Markus Jakobsson, Elaine Shi, Philippe Golle, and Richard Chow. Implicit Authentication for Mobile Devices. pages 9–9, 2009.

[180] H. K. Jayasuriya. *Big data, big challenges in evidence-based policymaking*. West Academic Publishing, St. Paul, MN, 2015.

[181] Jiong Jin, Jayavardhana Gubbi, Slaven Marusic, and Marimuthu Palaniswami. An information framework for creating a smart city through internet of things. *IEEE Internet of Things Journal*, 1(2):112–121, 2014.

[182] Jiong Jin, Jayavardhana Gubbi, Slaven Marusic, and Marimuthu Palaniswami. An information framework for creating a smart city through internet of things. *IEEE Internet of Things Journal*, 1(2):112–121, 2014.

[183] Felice Armenio, Mark Harrison, Bernie Hogan, Jin Mitsugi, Josef Preishuber, Oleg Ryaboy, and KK Suen. The epcglobal architecture framework. 2005.

[184] Rohan Joshi, Deedee Kommers, Laurien Oosterwijk, Loe Feijs, Carola Van Pul, and Peter Andriessen. Predicting Neonatal Sepsis Using Features of Heart Rate Variability, Respiratory Characteristics and ECG-Derived Estimates of Infant Motion. *IEEE Journal of Biomedical and Health Informatics*, pages 1–1, 2019.

[185] M. Kalmeshwar and N. Prasad. Internet of things: architecture, issues and applications. *Int. J. Eng. Res. Appl*, 7(06):85–88, 2017.

[186] Hazalila Kamaludin, Hairulnizam Mahdin, and Jemal H Abawajy. Clone tag detection in distributed rfid systems. *PloS One*, 13(3):e0193951, 2018.

[187] A Kannammal, et al. Security attacks and defensive measures for routing mechanisms in manets–a survey. *International Journal of Computer Applications*, 42(4):27–32, 2012.

[188] A. Kapsalis, P. Kasnesis, I. S. Venieris, D. I. Kaklamani, and C. Z. Patrikakis. A cooperative fog approach for effective workload balancing. *IEEE Cloud Computing*, 4(2):36–45, 2017.

[189] Abdul Karim, Avinash Mishra, M.A. Hakim Newton, and Abdul Sattar. Machine learning interpretability: A science rather than a tool, 2018.

[190] N. Karthik and V.S. Ananthanarayana. Context aware trust management scheme for pervasive healthcare. *Wireless Personal Communications*, 105(3):725–763, 2019.

[191] Rahat Ali Khan and Al-Sakib Khan Pathan. The state-of-the-art wireless body area sensor networks: A survey. *International Journal of Distributed Sensor Networks*, 14(4):1550147718768994, 2018.

[192] Jihoon Kim, Hyeoneui Kim, Elizabeth Bell, Tyler Bath, Paulina Paul, Anh Pham, Xiaoqian Jiang, Kai Zheng, and Lucila Ohno-Machado. Patient Perspectives About Decisions to Share Medical Data and Biospecimens for Research. *JAMA Network Open*, 2(8), August 2019.

[193] J.S. Kiran, M. Sravanthi, K. Preethi, M. Anusha. Recent Issues and Challenges on Big Data in Cloud Computing. *Int. J. of Computer Science and Information Security (IJCSIS)*, 6:98–102, 2015.

[194] J. Kunzi, R. P. (Paul) Koster, and M. Petkovic. Emergency Access to Protected Health Records. *Medical Informatics in a United and Healthy Europe (Proceedings of MIE 2009, The XXIInd International Congress of the European Federation for Medical Informatics, Sarajevo, Bosnia and Herzegovina, August 30-September 2, 2009)*, pages 705–709, 2009.

[195] J. Kobielus. *Deployment in Big Data Analytics Applications to the Cloud: Road Map for Success*. Cloud Standards Customer Council, Technical report, 2014.

[196] Marko Kompara and Marko Holbl. Survey on security in intra-body area network communication. *Ad Hoc Networks*, 70:23–43, 2018.

[197] Jakub Konečný, H. Brendan McMahan, Felix X. Yu, Peter Richtárik, Ananda Theertha Suresh, and Dave Bacon. Federated learning: Strategies for improving communication efficiency. *arXiv preprint arXiv:1610.05492*, 2016.

[198] Teemu Koponen, Mohit Chawla, Byung-Gon Chun, Andrey Ermolinskiy, Kye Hyun Kim, Scott Shenker, and Ion Stoica. A data-oriented (and beyond) network architecture. *SIGCOMM Comput. Commun. Rev.*, 37(4):181–192, August 2007.

[199] Patty Kostkova, Helen Brewer, Simon Lusignan, and Edward Fottrell. Who Owns the Data? Open Data for Healthcare. *Frontiers Public Health*, 4.7, February 2016.

[200] Djamel Eddine Kouicem, Abdelmadjid Bouabdallah, and Hicham Lakhlef. Internet of things security: A top-down survey. *Computer Networks*, 141:199–221, 2018.

[201] Y. N. Krishnan, C. N. Bhagwat, and A. P. Utpat. Fog computing - network based cloud computing. In *2015 2nd International Conference on Electronics and Communication Systems (ICECS)*, pages 250–251, 2015.

[202] Katharina Krombholz, Heidelinde Hobel, Markus Huber, and Edgar Weippl. Social engineering attacks on the knowledge worker. In *ACM SIN '13*, pages 28–35, Aksaray, Turkey, November 2013. ACM.

[203] Clemens Scott Kruse, Anna Stein, Heather Thomas, and Harmander Kaur. The Use of Electronic Health Records to Support Population Health: A Systematic Review of the Literature. *Journal of Medical Systems*, 42(11), 2018.

[204] Pallavi Kulkarni and Rajashri Khanai. Addressing mobile cloud computing security issues: A survey. In *IEEE ICCSP 2015 Conference*, pages 1463–1467, Bangalore, India, 2015.

[205] Mascha Kurpicz, Anne-Cécile Orgerie, and Anita Sobe. How much does a vm cost? Energy-proportional accounting in vm-based environments. In *2016 24th Euromicro International Conference on Parallel, Distributed, and Network-Based Processing (PDP)*, pages 651–658. IEEE, 2016.

[206] D. Kutscher, S. Eum, K. Pentikousis, I. Psaras, D. Corujo, D. Saucez, T. Schmidt, and M. Waehlisch. Information-centric networking (icn) research challenges. *RFC 7927*, (1), July 2016.

[207] Dimosthenis Kyriazis, Theodora Varvarigou, Daniel White, Andrea Rossi, and Joshua Cooper. Sustainable smart city iot applications: Heat and electricity management & eco-conscious cruise control for public transportation. In *2013 IEEE 14th International Symposium on "A World of Wireless, Mobile and Multimedia Networks"(WoWMoM)*, pages 1–5. IEEE, 2013.

[208] Melinda K. Lacy, Neil E. Klutman, Rebecca T. Horvat, and Antonia Zapantis. Antibiograms: New NCCLS Guidelines, Development, and Clinical Application. *Hospital Pharmacy*, 39(6):542–553, June 2004.

[209] Xiaochen Lai, Quanli Liu, Xin Wei, Wei Wang, Guoqiao Zhou, and Guangyi Han. A survey of body sensor networks. *Sensors*, 13(5):5406–5447, 2013.

[210] Eric J. Lammers, Julia Adler-Milstein, and Keith E. Kocher. Does Health Information Exchange Reduce Redundant Imaging? Evidence From Emergency Departments. *Medical Care*, 52(3):227, March 2014.

[211] Robert Langreth and Michael Waldholz. New Era of Personalized Medicine Targeting Drugs For Each Unique Genetic Profile. *The Oncologist*, 4(5):426–427, October 1999.

[212] Rabia Latif, Haider Abbas, and Saïd Assar. Distributed denial of service (DDoS) attack in cloud-assisted wireless body area networks: A systematic literature review. *Journal of Medical Systems*, 38(11):128, 2014.

[213] Benoît Latré, Bart Braem, Ingrid Moerman, Chris Blondia, and Piet Demeester. A survey on wireless body area networks. *Wirel. Netw.*, 17(1):1–18, January 2011.

[214] Teemu Leppanen, Jose Alvarez Lacasia, Yoshito Tobe, Kaoru Sezaki, and Jukka Riekki. Mobile crowdsensing with mobile agents: Jaamas extended abstract. In *Proceedings of the 2016 International Conference on Autonomous Agents & Multiagent Systems*, pages 738–739. International Foundation for Autonomous Agents and Multiagent Systems, 2016.

[215] C. Levy-Bencheton and E. Darra. Cyber security and resilience of intelligent public transport: Good practices and recommendations. 2015.

[216] Chao Li, Zhenjiang Zhang, Zhangbing Zhou, Yingsi Zhao, and Naiyue Chen. A UKF-based emergency aware fusion model in a heterogeneous network for wireless body networks. *IEEE Access*, 2019.

[217] Chun-Ta Li, Cheng-Chi Lee, and Chi-Yao Weng. A secure cloud-assisted wireless body area network in mobile emergency medical care system. *Journal of Medical Systems*, 40(5):117, 2016.

[218] J. Li, J. Jin, D. Yuan, M. Palaniswami, and K. Moessner. Ehopes: Data-centered fog platform for smart living. In *2015 International Telecommunication Networks and Applications Conference (ITNAC)*, 2015.

[219] M. Li, S. Yu, Y. Zheng, K. Ren, and W. Lou. Scalable and Secure Sharing of Personal Health Records in Cloud Computing Using Attribute-Based Encryption. *IEEE Transactions on Parallel and Distributed Systems*, 24(1):131–143, January 2013.

[220] Ming Li, Shucheng Yu, Kui Ren, and Wenjing Lou. Securing Personal Health Records in Cloud Computing: Patient-Centric and Fine-Grained Data Access Control in Multi-owner Settings. In Sushil Jajodia and Jianying Zhou, editors, *Security and Privacy in Communication Networks*, pages 89–106. Springer, Berlin Heidelberg, 2010.

[221] S. Li, Y. Zhang, D. Raychaudhuri, and R. Ravindran. A comparative study of MobilityFirst and ndn based icn-iot architectures. In *10th International Conference on Heterogeneous Networking for Quality, Reliability, Security and Robustness*, pages 158–163, Aug 2014.

[222] Shancang Li. Security requirements in iot architecture. *Securing the Internet of Things*, pages 97–108, 2017.

[223] Xiaohui Liang, Rongxing Lu, Le Chen, Xiaodong Lin, and Xuemin Shen. PEC: A privacy-preserving emergency call scheme for mobile healthcare social networks. *Journal of Communications and Networks*, 13(2):102–112, 2011.

[224] Elizabeth O. Lillie, Bradley Patay, Joel Diamant, Brian Issell, Eric J. Topol, and Nicholas J. Schork. The N-of-1 Clinical Trial: The Ultimate Strategy for Individualizing Medicine?, March 2011.

[225] H. Lin, J. Shao, C. Zhang, and Y. Fang. CAM: Cloud-assisted privacy preserving mobile health monitoring. *IEEE Transactions on Information Forensics and Security*, 8(6):985–997, June 2013.

[226] Huichen Lin and Neil Bergmann. Iot privacy and security challenges for smart home environments. *Information*, 7(3):44, 2016.

[227] Jie Lin, Wei Yu, Nan Zhang, Xinyu Yang, Hanlin Zhang, and Wei Zhao. A survey on internet of things: Architecture, enabling technologies, security and privacy, and applications. *IEEE Internet of Things Journal*, 4(5):1125–1142, 2017.

[228] P.J.G. Lisboa. A review of evidence of health benefit from artificial neural networks in medical intervention. *Neural Networks*, 15(1):11 – 39, 2002.

[229] Jingwei Liu, Zonghua Zhang, Xiaofeng Chen, and Kyung Sup Kwak. Certificateless remote anonymous authentication schemes for wirelessbody area networks. *IEEE Transactions on Parallel and Distributed Systems*, 25(2):332–342, 2013.

[230] Ryoo J. Liu, Y. and S. Rizvi. In *Proceedings of IEEE 23rd Wireless and Optimal Communication Conference (WOCC)*, pages 1–6.

[231] Chi-Chun Lo, Chun-Chieh Huang, and Joy Ku. Cooperative intrusion detection system framework for cloud computing networks. In *First IEEE International Conference on Ubi-Media Computing*, pages 280–284. IEEE, 2008.

[232] Steve Lohr. Google Offers Personal Health Records on the Web. *The New York Times*, May 2008.

[233] Steve Lohr. Google Is Closing Its Health Records Service. *The New York Times*, June 2011.

[234] Eva Lonn, Jackie Bosch, Koon K. Teo, Prem Pais, Denis Xavier, and Yusuf Salim. The Polypill in the Prevention of Cardiovascular Diseases. *Circulation*, 122(20):2078–2088, November 2010.

[235] Diego M. Lopez and Bernd G. M. E. Blobel. A Development Framework for Semantically Interoperable Health Information Systems. *International Journal of Medical Informatics*, 78(2):83–103, February 2009.

[236] Rongxing Lu, Xiaodong Lin, Xiaohui Liang, and Xuemin Shen. A secure handshake scheme with symptoms-matching for mhealthcare social network. *Mobile Networks and Applications*, 16(6):683–694, 2011.

[237] David Luxton, Robert Kayl, and Matthew Mishkind. mHealth Data Security: The Need for HIPAA-Compliant Standardization. *Telemedicine Journal and e-health : The Official Journal of the American Telemedicine Association*, 18:284–8, 2012.

[238] Michael Mackay, Thar Baker, and Adil Al-Yasiri. Security-oriented cloud computing platform for critical infrastructures. *Computer Law & Security Review*, 28:679–686, 2012.

[239] Somayya Madakam, R. Ramaswamy, and Siddharth Tripathi. Internet of things (iot): A literature review. *Journal of Computer and Communications*, 3(05):164, 2015.

[240] Jeanne M. Madden, Matthew D. Lakoma, Donna Rusinak, Christine Y. Lu, and Stephen B. Soumerai. Missing clinical and behavioral health data in a large electronic health record (EHR) system. *Journal of the American Medical Informatics Association*, 23(6):1143–1149, November 2016.

[241] Aravindh Mahendran and Andrea Vedaldi. Understanding deep image representations by inverting them. In *Proceedings of the IEEE Conference on Computer Vision and Pattern Recognition*, pages 5188–5196, 2015.

[242] Ahmed Redha Mahlous. Internet of things (iots): Architecture, evolution, threats and defense. *International Journal of Computer Science and Network Security*, 19(1).

[243] T. Maitra and S. Roy. Research challenges in BAN due to the mixed WSN features: Some perspectives and future directions. *IEEE Sensors Journal*, 17(17):5759–5766, Sep. 2017.

[244] M. F. Majeed, S. H. Ahmed, and M. N. Dailey. Enabling push-based critical data forwarding in vehicular named data networks. *IEEE Communications Letters*, 21(4):873–876, April 2017.

[245] Jacqueline Major, M. Oliver, Chyke Doubeni, Albert Hollenbeck, B. Graubard, and Rajiv Sinha. Re: Socioeconomic status, healthcare density, and risk of prostate cancer among African American and caucasian

men in a large prospective study. *The Journal of Urology*, 189:2108–2108, 06 2013.

[246] Donna Malvey and Donna Slovensky. *mHealth: Transforming Healthcare.* Springer Publishing Company, Incorporated, 2014.

[247] Kenneth D. Mandl, William W. Simons, William C.R. Crawford, and Jonathan M. Abbett. Indivo: A personally controlled health record for health information exchange and communication. *BMC Medical Informatics and Decision Making*, 7(1):25, September 2007.

[248] Avinash Chandra Mani and Anil Kumar Malviya. Security challenges in cloud computing networks. *Available at SSRN 3350319*, 2019.

[249] J. Mann. The internet of things: Opportunities and applications across industries. In *International Institute for Analytics*, 2015.

[250] Gunasekaran Manogaran, R. Varatharajan, Daphne Lopez, Priyan Malarvizhi Kumar, Revathi Sundarasekar, and Chandu Thota. A new architecture of internet of things and big data ecosystem for secured smart healthcare monitoring and alerting system. *Future Generation Comp. Syst.*, 82:375–387, 2017.

[251] M.S. Marcolino, J.A.Q. Oliveira, and M. D'Agostino. The Impact of mHealth Interventions: Systematic Review of Systematic Reviews. *JMIR Mhealth Uhealth*, 6:e23, January 2018.

[252] Izet Masic, Milan Miokovic, and Belma Muhamedagic. Evidence Based Medicine - New Approaches and Challenges. *Acta Informatica Medica*, 16(4):219–225, 2008.

[253] Michael Mcgrath and Cliodhna Ni Scanaill. *Sensor technologies: Healthcare, wellness, and environmental applications.* 01 2013.

[254] Brendan McMahan and Daniel Ramage. Federated learning: Collaborative machine learning without centralized training data. *Google Research Blog*, 3, 2017.

[255] H. Brendan McMahan, Eider Moore, Daniel Ramage, Seth Hampson, et al. Communication-efficient learning of deep networks from decentralized data. *arXiv preprint arXiv:1602.05629*, 2016.

[256] Barbara McPake and Ajay Mahal. Addressing the needs of an aging population in the health system: The Australian case. *Health Systems & Reform*, 3(3):236–247, 2017. PMID: 31514669.

[257] Patricia Mechael, Nadi Kaonga, and Hima Batavia. mHealth: New Horizons for Health Through Mobile Technologies. Technical report, World Health Organization, 2011.

[258] M. Meddeb, A. Dhraief, A. Belghith, T. Monteil, and K. Drira. Cache coherence in machine-to-machine information centric networks. In *2015 IEEE 40th Conference on Local Computer Networks (LCN)*, pages 430–433, Oct 2015.

[259] Maroua Meddeb, Amine Dhraief, Abdelfettah Belghith, Thierry Monteil, Khalil Drira, and Saad Al-Ahmadi. Named data networking: A promising architecture for the internet of things (iot). *International Journal on Semantic Web and Information Systems (IJSWIS)*, 14(2):86–112, 2018.

[260] Peter Mell, Tim Grance, et al. The nist definition of cloud computing. 2011.

[261] Diego M. Mendez, Ioannis Papapanagiotou, and Baijian Yang. Internet of things: Survey on security and privacy. *arXiv preprint arXiv:1707.01879*, 2017.

[262] Stephen T Mennemeyer, Nir Menachemi, Saurabh Rahurkar, and Eric W. Ford. Impact of the HITECH Act on physicians' adoption of electronic health records. *Journal of the American Medical Informatics Association*, 23(2):375–379, 07 2015.

[263] D. Metcalf, S. T. J. Milliard, M. Gomez, and M. Schwartz. Wearables and the Internet of Things for Health: Wearable, Interconnected Devices Promise More Efficient and Comprehensive Health Care. *IEEE Pulse*, 7(5):35–39, September 2016.

[264] Dragorad A. Milovanovic and Zoran S. Bojkovic. Cloud-based iot healthcare applications: Requirements and recommendations. *International Journal of Internet of Things and Web Services*, 2:60–65.

[265] Y. G. Min, H. J. Shin, and Y. H. Bang. Cloud Computing and Big Data: A Review of Current Service Models and Hardware Perspectives. *Journal of Engineering*, 9(2):686–693, 2012.

[266] Nateq Be-Nazir Ibn Minar and Mohammed Tarique. Bluetooth security threats and solutions: A survey. *International Journal of Distributed and Parallel Systems*, 3(1):127, 2012.

[267] Roberto Minerva, Abyi Biru, and Domenico Rotondi. Towards a definition of the internet of things (iot). *IEEE Internet Initiative*, 1:1–86, 2015.

[268] Chang Ming, Valeria Viassolo, Nicole Probst-Hensch, Pierre O. Chappuis, Ivo D. Dinov, and Maria C. Katapodi. Machine learning techniques for personalized breast cancer risk prediction: Comparison with the BCRAT and BOADICEA models. *Breast Cancer Research*, 21(1):75, June 2019.

[269] Daniel Minoli and Benedict Occhiogrosso. Blockchain mechanisms for iot security. *Internet of Things*, 1:1–13, 2018.

[270] Tanveer Ahmad Mir and Makoto Nakamura. 3d-bioprinting: Towards the era of manufacturing human organs as spare parts for healthcare and medicine. *Tissue Engineering Part B Reviews*, 23, 01 2017.

[271] Preeti Mishra, Emmanuel S. Pilli, Vijay Varadharajan, and Udaya Tupakula. Intrusion detection techniques in cloud environment: A survey. *Journal of Network and Computer Applications*, 77:18–47, 2017.

[272] Preeti Mishra, Emmanuel S. Pilli, Vijay Varadharajan, and Udaya Tupakula. Psi-netvisor: Program semantic aware intrusion detection at network and hypervisor layer in cloud. *Journal of Intelligent & Fuzzy Systems*, 32(4):2909–2921, 2017.

[273] Chirag Modi, Dhiren Patel, Bhavesh Borisaniya, Avi Patel, and Muttukrishnan Rajarajan. A survey on security issues and solutions at different layers of cloud computing. *Journal of Supercomputing*, 63:561–592, 2013.

[274] Shahriar Mohammadi and Hossein Jadidoleslamy. A comparison of link layer attacks on wireless sensor networks. *arXiv preprint arXiv:1103.5589*, 2011.

[275] A. Monteiro, H. Dubey, L. Mahler, Q. Yang, and K. Mankodiya. Fit: A fog computing device for speech tele-treatments. In *2016 IEEE International Conference on Smart Computing (SMARTCOMP)*, pages 1–3, 2016.

[276] J. Morak, M. Schwarz, D. Hayn, and G. Schreier. Feasibility of mHealth and Near Field Communication Technology Based Medication Adherence Monitoring. In *2012 Annual International Conference of the IEEE Engineering in Medicine and Biology Society*, pages 272–275, August 2012.

[277] Arsalan Mosenia and Niraj K Jha. A comprehensive study of security of internet-of-things. *IEEE Transactions on Emerging Topics in Computing*, 5(4):586–602, 2016.

[278] Arsalan Mosenia, Susmita Sur-Kolay, Anand Raghunathan, and Niraj K. Jha. Wearable medical sensor-based system design: A survey. *IEEE Transactions on Multi-Scale Computing Systems*, 3(2):124–138, 2017.

[279] C. Mouradian, D. Naboulsi, S. Yangui, R. H. Glitho, M. J. Morrow, and P. A. Polakos. A comprehensive survey on fog computing: State-of-the-art and research challenges. *IEEE Communications Surveys Tutorials*, 20(1), 2018.

[280] C. Mouradian, D. Naboulsi, S. Yangui, R. H. Glitho, M. J. Morrow, and P. A. Polakos. A comprehensive survey on fog computing: State-of-the-art and research challenges. *IEEE Communications Surveys Tutorials*, 20(1):416–464, 2018.

[281] S. Movassaghi, M. Abolhasan, J. Lipman, D. Smith, and A. Jamalipour. Wireless body area networks: A survey. *IEEE Communications Surveys Tutorials*, 16(3):1658–1686, Third Quarter 2014.

[282] Bidyut Mukherjee, Songjie Wang, Wenyi Lu, Roshan Lal Neupane, Daniel Dunn, Yijie Ren, Qi Su, and Prasad Calyam. Flexible iot security middleware for end-to-end cloud–fog communication. *Future Generation Computer Systems*, 87:688–703, 2018.

[283] Kyle Murphy. Early Data Shows High Patient Satisfaction With Apple's PHR, January 2019.

[284] Arslan Musaddiq, Yousaf Bin Zikria, Oliver Hahm, Heejung Yu, Ali Kashif Bashir, and Sung Won Kim. A survey on resource management in IoT operating systems. *IEEE Access*, 6:8459–8482, 2018.

[285] M.A. Naeem, R. Ali, S.A. Nor, and S. Hassan. A periodic caching strategy solution for the smart city in the information-centric internet of things. *Sustainability*, 10(2576), 2018.

[286] Antonio L. Maia Neto, Artur L. F. Souza, Italo Cunha, Michele Nogueira, Ivan Oliveira Nunes, Leonardo Cotta, Nicolas Gentille, Antonio A. F. Loureiro, Diego F. Aranha, Harsh Kupwade Patil, and Leonardo B. Oliveira. AoT: Authentication and Access Control for the Entire IoT Device Life-Cycle. In *Proceedings of the 14th ACM Conference on Embedded Network Sensor Systems CD-ROM*, SenSys '16, pages 1–15, New York, NY, USA, 2016. ACM. event-place: Stanford, CA, USA.

[287] P. C. Neves, B. Schmerl, J. Bernardino, and J. Camara. Big data in cloud computing: Features and issues. In *International Conference on Internet of Things and Big Data*, 2016.

[288] H.P. Ng, S.H. Ong, K.W.C. Foong, P.S. Goh, and W.L. Nowinski. Medical image segmentation using K-Means clustering and improved watershed algorithm. In *2006 IEEE Southwest Symposium on Image Analysis and Interpretation*, pages 61–65. IEEE, 2006.

[289] Arsalan Mohsen Nia, Susmita Sur-Kolay, Anand Raghunathan, and Niraj K Jha. Physiological information leakage: A new frontier in health information security. *IEEE Transactions on Emerging Topics in Computing*, 4(3):321–334, 2015.

[290] V.K. Nigam and S. Bhatia. *Impact of cloud computing on Healthcare, Cloud Standards Customer Service*. Technical report, 2014.

[291] Priya Oberoi and Sumit Mittal. Survey of various security attacks in clouds based environments. *International Journal of Advanced Research in Computer Science*, 8(9), 2017.

[292] Office of the National Coordinator for Health Information Technology. Non-federal Acute Care Hospital Electronic Health Record Adoption.

[293] Folasade Osisanwo, Shade Kuyoro, and Oludele Awodele. Internet refrigerator–a typical internet of things (iot). In *3rd International Conference on Advances in Engineering Sciences & Applied Mathematics (ICAESAM'2015), London (UK). http://iieng. org/images/proceedings_pdf/2602E0315051. pdf,* 2015.

[294] Rafail Ostrovsky, Amit Sahai, and Brent Waters. Attribute-based encryption with non-monotonic access structures. In *Proceedings of the 14th ACM Conference on Computer and Communications Security,* pages 195–203. ACM, 2007.

[295] Hamed Haddad Pajouh, Reza Javidan, Raouf Khayami, Dehghantanha Ali, and Kim-Kwang Raymond Choo. A two-layer dimension reduction and two-tier classification model for anomaly-based intrusion detection in iot backbone networks. *IEEE Transactions on Emerging Topics in Computing,* 2016.

[296] Harshal Pandve and Tukaram Pandve. Primary healthcare system in India: Evolution and challenges. *International Journal of Health System and Disaster Management,* 1(3):125–128, 2013.

[297] Dorottya Papp, Zhendong Ma, and Levente Buttyan. Embedded systems security: Threats, vulnerabilities, and attack taxonomy. In *2015 13th Annual Conference on Privacy, Security and Trust (PST),* pages 145–152. IEEE, 2015.

[298] Kihong Park and Heejo Lee. On the effectiveness of route-based packet filtering for distributed dos attack prevention in power-law internets. In *ACM SIGCOMM Computer Communication Review,* volume 31, pages 15–26. ACM, 2001.

[299] Roy C. Park, Hoill Jung, Dong-Kun Shin, Gui-Jung Kim, and Kun-Ho Yoon. M2m-Based Smart Health Service for Human UI/UX Using Motion Recognition. *Cluster Computing,* 18(1):221–232, March 2015.

[300] Isha Pathak and Deo Prakash Vidyarthi. A model for virtual network embedding across multiple infrastructure providers using genetic algorithm. *Science China Information Sciences,* 2017.

[301] Anupam Pattanayak. Revisiting dedicated and block cipher based hash functions. *IACR Cryptology ePrint Archive,* 2012:322, 2012.

[302] Anupam Pattanayak and Banshidhar Majhi. Key predistribution schemes in distributed wireless sensor network using combinatorial designs revisited. *IACR Cryptology ePrint Archive,* 2009:131, 2009.

[303] Pablo Punal Pereira, Jens Eliasson, and Jerker Delsing. An authentication and access control framework for coap-based internet of things. In *IECON 2014-40th Annual Conference of the IEEE Industrial Electronics Society*, pages 5293–5299. IEEE, 2014.

[304] Diego Perez-Botero, Jakub Szefer, and Ruby Lee. Characterizing hypervisor vulnerabilities in cloud computing servers. In *CloudComputing'13*, pages 3–10, Hangzhou, China, May 2013. ACM.

[305] Francesca Pillemer, Rebecca Anhang Price, Suzanne Paone, G. Daniel Martich, Steve Albert, Leila Haidari, Glenn Updike, Robert Rudin, Darren Liu, and Ateev Mehrotra. Direct Release of Test Results to Patients Increases Patient Engagement and Utilization of Care. *PLOS ONE*, 11(6):e0154743, June 2016.

[306] Pavan Pongle and Gurunath Chavan. A survey: Attacks on rpl and 6lowpan in iot. In *2015 International Conference on Pervasive Computing (ICPC)*, pages 1–6. IEEE, 2015.

[307] G. Premsankar, M. Di Francesco, and T. Taleb. Edge computing for the internet of things: A case study. *IEEE Internet of Things Journal*, 5(2):1275–1284, April 2018.

[308] Peter J. Pronovost, Deborah Baugher Hobson, Karen A. Earsing, Elizabeth S. Lins, Michael L. Rinke, Katherine Emery, Sean M. Berenholtz, Pamela A. Lipsett, and Todd Dorman. A Practical Tool to Reduce Medication Errors During Patient Transfer from an Intensive Care Unit. 2004.

[309] Deepak Puthal, Surya Nepal, Rajiv Ranjan, and Jinjun Chen. Threats to networking cloud and edge datacenters in the internet of things. *IEEE Cloud Computing*, 3(3):64–71, 2016.

[310] M. G. Rabbani, R. P. Esteves, M. Podlesny, G. Simon, L. Z. Granville, and R. Boutaba. On tackling virtual data center embedding problem. In *2013 IFIP/IEEE International Symposium on Integrated Network Management (IM 2013)*, pages 177–184, 2013.

[311] R. Ravindran, Xuan Liu, A. Chakraborti, Xinwen Zhang, and Guoqiang Wang. Towards software defined icn based edge-cloud services. In *2013 IEEE 2nd International Conference on Cloud Networking (CloudNet)*, pages 227–235, Nov 2013.

[312] R. Ravindran, Y. Zhang, L. Grieco, A. Cindgren, D. Raychadhuri, E. Baccelli, J. Burke, G. Wang, B. Ahlgren, and O. Schelen. Design considerations for applying icn to iot. *ICN Research Group*, February 2018.

[313] M. A. Razzaq, S. H. Gill, M. A. Qureshi, and S. Ullah. Security Issues in the Internet of Things (IoT): A Comprehensive Study. *Int. J. of computer Science and Information Security (IJCSIS)*, 8(6), 2017.

[314] Irina Rish. In *IJCAI 2001 workshop on empirical methods in artificial intelligence*, 2001.

[315] John C. Roberts and Wasim Al-Hamdani. Who can you trust in the cloud? A review of security issues within cloud computing. In *Information Security Curriculum Development Conference*. ACM, 2011.

[316] Patrick Ryan and Sarah Falvey. Trust in the clouds. *Computer Law & Security Review*, 28:513–521, 2012.

[317] David L. Sackett. Evidence-based medicine. *Seminars in Perinatology*, 21(1):3–5, February 1997.

[318] David L. Sackett, William M. C. Rosenberg, J. A. Muir Gray, R. Brian Haynes, and W. Scott Richardson. Evidence based medicine: What it is and what it isn't. *BMJ*, 312(7023):71–72, January 1996.

[319] Yuvraj Sahni, Jiannong Cao, Shigeng Zhang, and Lei Yang. Edge mesh: A new paradigm to enable distributed intelligence in internet of things. *IEEE access*, 5:16441–16458, 2017.

[320] Anam Sajid and Haider Abbas. Data privacy in cloud-assisted healthcare systems: State of the art and future challenges. *Journal of Medical Systems*, 40(6):155, 2016.

[321] Tomas Sanchez, DC Ranasinghe, Mark Harrison, and Duncan McFarlane. Adding sense to the internet of things-An architecture framework for smart object systems. *Pers Ubiquitous Comput*, 16(3):291–308, 2012.

[322] R. Saranya and V.P. Muthukumar. Security Issues Associated with Big Data in Cloud Computing. *Int. J. of Multidisciplinary and Development*, 2:580–585.

[323] Anurag Satpathy, Sourav Kanti Addya, Ashok Kumar Turuk, Banshidhar Majhi, and Gadadhar Sahoo. Crow search based virtual machine placement strategy in cloud data centers with live migration. *Computers and Electrical Engineering*, 69:334–350, 2018.

[324] Arbia Riahi Sfar, Zied Chtourou, and Yacine Challal. A systemic and cognitive vision for iot security: A case study of military live simulation and security challenges. *International Conference on Smart, Monitored and Controlled Cities*.

[325] Hossein Shafagh, Lukas Burkhalter, Anwar Hithnawi, and Simon Duquennoy. Towards Blockchain-based Auditable Storage and Sharing of IoT Data. In *Proceedings of the 2017 on Cloud Computing Security Workshop - CCSW '17*, pages 45–50, Dallas, Texas, USA, 2017. ACM Press.

[326] W. Shang, Q. Ding, A. Marianantoni, J. Burke, and L. Zhang. Securing building management systems using named data networking. *IEEE Network*, 28(3):50–56, May 2014.

[327] W. Shang, Y.Yu, R.Droms, and L.Zhang. Challenges in iot networking via tcp/ip architecture. In *NDN Project*, 2016.

[328] Elaine Shi, Yuan Niu, Markus Jakobsson, and Richard Chow. Implicit Authentication through Learning User Behavior. In Mike Burmester, Gene Tsudik, Spyros Magliveras, and Ivana Ilić, editors, *Information Security*, pages 99–113. Springer, Berlin Heidelberg, 2011.

[329] H. Shi, N. Chen, and R. Deters. Combining mobile and fog computing: Using coap to link mobile device clouds with fog computing. In *2015 IEEE International Conference on Data Science and Data Intensive Systems*, pages 564–571, 2015.

[330] R. Silva, J. S. Silva, and F. Boavida. Infrastructure-supported mobility in wireless sensor networks - a case study. In *2015 IEEE International Conference on Industrial Technology (ICIT)*, pages 1895–1900, March 2015.

[331] M. Simon. Internet of things security issues bleed into 2018. In *https://www.helpnetsecurity.com/2018/01/16/internet-of-things-security-issues-2018/*, 2018.

[332] William W. Simons, Kenneth D. Mandl, and Isaac S. Kohane. The PING Personally Controlled Electronic Medical Record System: Technical Architecture. *Journal of the American Medical Informatics Association : JAMIA*, 12(1):47–54, 2005.

[333] Karen Simonyan and Andrew Zisserman. Very deep convolutional networks for large-scale image recognition, 2014.

[334] Smita Singh, Sarita Negi, Shashi Kant Verma, and Neelam Panwar. Comparative study of existing data scheduling approaches and role of cloud in vanet environment. *Procedia Computer Science*, 125:925–934, 2018.

[335] M. Slabicki and K. Grochla. Performance evaluation of coap, snmp and netconf protocols in fog computing architecture. In *NOMS 2016 - 2016 IEEE/IFIP Network Operations and Management Symposium*, pages 1315–1319, 2016.

[336] Bongjin Sohn, Sungpil Woo, Jiyong Han, Hyeeun Cho, Jaewook Byun, and Daeyoung Kim. Gs1 connected car using epcis-ons system. In *2016 IEEE International Congress on Big Data (BigData Congress)*, pages 426–429. IEEE, 2016.

[337] Madhan Kumar Srinivasan, K Sarukesi, Paul Rodrigues, Sai Manoj, and Revathy. State-of-the-art cloud computing security taxonomies – a classification of security challenges in the present cloud computing environment. In *ICACCI '12 Proceedings of the International Conference on Advances in Computing, Communications and Informatics*, pages 470–476. ACM, 2012.

[338] Vladimir Stantchev, Ahmed Barnawi, Sarfaraz Ghulam, Johannes Schubert, and Gerrit Tamm. Smart items, fog and cloud computing as enablers of servitization in healthcare. 2015.

[339] Robert Steinbrook. Personally Controlled Online Health Data - The Next Big Thing in Medical Care? *New England Journal of Medicine*, 358(16):1653–1656, April 2008.

[340] Nalini Subramanian and Andrews Jeyaraj. Recent security challenges in cloud computing. *Computers & Electrical Engineering*, 71:28–42.

[341] Gang Sun, Dan Liao, Sitong Bu, Hongfang Yu, Zhili Sun, and Victor Chang. The efficient framework and algorithm for provisioning evolving vdc in federated data centers. *Future Generation Computer Systems*, 73:79 – 89, 2017.

[342] Gang Sun, Dan Liao, Dongcheng Zhao, Zhili Sun, and Victor Chang. Towards provisioning hybrid virtual networks in federated cloud data centers. *Future Generation Computer Systems*, 87:457 – 469, 2018.

[343] Gang Sun, Hongfang Yu, Vishal Anand, and Lemin Li. A cost efficient framework and algorithm for embedding dynamic virtual network requests. *Future Generation Computer Systems*, 29(5):1265 – 1277, 2013.

[344] J. Sun and Y. Fang. Cross-Domain Data Sharing in Distributed Electronic Health Record Systems. *IEEE Transactions on Parallel and Distributed Systems*, 21(6):754–764, June 2010.

[345] Jinyuan Sun, Yuguang Fang, and Xiaoyan Zhu. Privacy and emergency response in e-healthcare leveraging wireless body sensor networks. *IEEE Wireless Communications*, 17(1):66–73, 2010.

[346] Peng Gang Sun, Lin Gao, and Shan Han. Prediction of human disease-related gene clusters by clustering analysis. *International Journal of Biological Sciences*, 7(1):61, 2011.

[347] P. Sundaravadivel, E. Kougianos, S. P. Mohanty, and M. K. Ganapathiraju. Everything you wanted to know about smart health care: Evaluating the different technologies and components of the internet of things for better health. *IEEE Consumer Electronics Magazine*, 7(1):18–28, Jan 2018.

[348] Hui Suo, Jiafu Wan, Caifeng Zou, and Jianqi Liu. Security in the internet of things: a review. In *2012 International Conference on Computer Science and Electronics Engineering*, volume 3, pages 648–651. IEEE, 2012.

[349] Roger L. Sur and Philipp Dahm. History of Evidence-Based Medicine. *Indian Journal of Urology : IJU : Journal of the Urological Society of India*, 27(4):487–489, 2011.

[350] Melanie Swan. Emerging Patient-Driven Health Care Models: An Examination of Health Social Networks, Consumer Personalized Medicine and Quantified Self-Tracking. *International Journal of Environmental Research and Public Health*, 6(2):492–525, February 2009.

[351] M. Swarna, S. Ravi, and M. Anand. Leaky bucket algorithm for congestion control. *International Journal of Applied Engineering Research*, 11(5):3155–3159, 2016.

[352] Yar-Ling Tan, Bok-Min Goi, Ryoichi Komiya, and Syh-Yuan Tan. A study of attribute-based encryption for body sensor networks. In *International Conference on Informatics Engineering and Information Science*, pages 238–247. Springer, 2011.

[353] Shinsuke Tanaka, Kenzaburo Fujishima, Nodoka Mimura, Dr. Eng., Tetsuya Ohashi, and Mayuko Tanaka. IoT Security IoT System Security Issues and Solution Approaches. *Hitachi Review*, 65(8):359–363, 2015.

[354] Hidema Tanaka. Evaluation of information leakage via electromagnetic emanation and effectiveness of tempest. *IEICE Transactions on Information and Systems*, 91(5):1439–1446, 2008.

[355] Bo Tang, Zhen Chen, Gerald Hefferman, Tao Wei, Haibo He, and Qing Yang. A hierarchical distributed fog computing architecture for big data analysis in smart cities. In *Proceedings of the ASE BigData & SocialInformatics 2015*, ASE BD&SI '15, 2015.

[356] L. M. R. Tarouco, L. M. Bertholdo, L. Z. Granville, L. M. R. Arbiza, F. Carbone, M. Marotta, and J. J. C. de Santanna. Internet of Things in healthcare: Interoperatibility and security issues. In *2012 IEEE International Conference on Communications (ICC)*, pages 6121–6125, June 2012.

[357] Kyle Taylor and Laura Silver. Smartphone Ownership Is Growing Rapidly Around the World, but Not Always Equally. Technical report, Pew Research Center, February 2019.

[358] Paul D. Thompson, Gregory Panza, Amanda Zaleski, and Beth Taylor. Statin-Associated Side Effects. *Journal of the American College of Cardiology*, 67(20):2395–2410, May 2016.

[359] Mark Tomlinson, Wesley Solomon, Yages Singh, Tanya Doherty, Mickey Chopra, Petrida Ijumba, Alexander C. Tsai, and Debra Jackson. The Use of Mobile Phones as a Data Collection Tool: A Report from a Household Survey in South Africa. *BMC Medical Informatics and Decision Making*, 9(1):51, December 2009.

[360] Yue Tong, Jinyuan Sun, Sherman S. M. Chow, and Pan Li. Cloud-assisted mobile-access of health data with privacy and auditability. *IEEE Journal of Biomedical and Health Informatics*, 18(2):419–429, 2013.

[361] Anne Townsend, Shelin Adam, Patricia H. Birch, Zoe Lohn, Francois Rousseau, and Jan M. Friedman. "I want to know what's in Pandora's box": Comparing stakeholder perspectives on incidental findings in clinical whole genomic sequencing. *American Journal of Medical Genetics Part A*, 158A(10):2519–2525, October 2012.

[362] N. B. Truong, G. M. Lee, and Y. Ghamri-Doudane. Software defined networking-based vehicular adhoc network with fog computing. In *2015 IFIP/IEEE International Symposium on Integrated Network Management (IM)*, 2015.

[363] Raylin Tso, Abdulhameed Alelaiwi, Sk Md Mizanur Rahman, Mu-En Wu, and M. Shamim Hossain. Privacy-preserving data communication through secure multi-party computation in healthcare sensor cloud. *Journal of Signal Processing Systems*, 89(1):51–59, 2017.

[364] Arijit Ukil, Jaydip Sen, and Sripad Koilakonda. Embedded security for internet of things. In *2011 2nd National Conference on Emerging Trends and Applications in Computer Science*, pages 1–6. IEEE, 2011.

[365] Dmitry Ulyanov, Andrea Vedaldi, and Victor Lempitsky. Deep image prior, 2017.

[366] Rohit Vaid and Vijay Katiyar. Security issues and remedies in wireless sensor networks-a survey. *International Journal of Computer Applications*, 79(4), 2013.

[367] Luis M. Vaquero, Luis Rodero-Merino, and Daniel Moron. Locking the sky: A survey on iaas cloud security. *Computing*, 91(1):93–118, 2011.

[368] Subramaniam Venkatraman, Kevin Pu Weekly, and Shelten Gee Jao Yuen. Method and apparatus for off-body detection for wearable device, June 2016.

[369] Joshua R. Vest and Larry D. Gamm. Health information exchange: Persistent challenges and new strategies. *Journal of the American Medical Informatics Association*, 17(3):288–294, May 2010.

[370] Wolfgang Vogt and Dorothea Nagel. Cluster analysis in diagnosis. *Clinical Chemistry*, 38(2):182–198, 1992.

[371] Ellen M. Voorhees and Narendra K. Gupta. Facilitating world wide web searches utilizing a multiple search engine query clustering fusion strategy, January 26 1999. US Patent 5,864,845.

[372] L. Wang, A.K.M. Mahmudul Hoque, and Cheng. Yi. Ospfn: An ospf based routing protocol for named data networking. *NDN, Technical Report NDN-0003*, July 2012.

[373] X. Wang, Q. Gui, B. Liu, Z. Jin, and Y. Chen. Enabling smart personalized healthcare: A hybrid mobile-cloud approach for ecg telemonitoring. *IEEE Journal of Biomedical and Health Informatics*, 18(3):739–745, May 2014.

[374] Robyn Whittaker, Hayden McRobbie, Chris Bullen, Anthony Rodgers, and Yulong Gu. Mobile Phone Based Interventions for Smoking Cessation. *Cochrane Database of Systematic Reviews*, (4), 2016.

[375] Ruoyu Wu, Gail-Joon Ahn, and Hongxin Hu. Towards hipaa-compliant healthcare systems. In *IHI'12 - Proceedings of the 2nd ACM SIGHIT International Health Informatics Symposium*, pages 593–601, 2012.

[376] Wanmin Wu, Sanjoy Dasgupta, Ernesto E. Ramirez, Carlyn Peterson, and Gregory J. Norman. Classification Accuracies of Physical Activities Using Smartphone Motion Sensors. *Journal of Medical Internet Research*, 14(5):e130, 2012.

[377] Hu Xiong. Cost-effective scalable and anonymous certificateless remote authentication protocol. *IEEE Transactions on Information Forensics and Security*, 9(12):2327–2339, 2014.

[378] Hu Xiong and Zhiguang Qin. Revocable and scalable certificateless remote authentication protocol with anonymity for wireless body area networks. *IEEE Transactions on Information Forensics and Security*, 10(7):1442–1455, 2015.

[379] G. Xylomenos, C. N. Ververidis, V. A. Siris, N. Fotiou, C. Tsilopoulos, X. Vasilakos, K. V. Katsaros, and G. C. Polyzos. A survey of information-centric networking research. *IEEE Communications Surveys Tutorials*, 16(2):1024–1049, Second 2014.

[380] Y. Yan and W. Su. A fog computing solution for advanced metering infrastructure. In *2016 IEEE/PES Transmission and Distribution Conference and Exposition (T D)*, 2016.

[381] Che-Ming Yang, Herng-Ching Lin, Polun Chang, and Wen-Shan Jian. Taiwan's perspective on electronic medical records' security and privacy protection: Lessons learned from HIPAA. *Computer Methods and Programs in Biomedicine*, 82(3):277–282, June 2006.

[382] Heetae Yang, Wonji Lee, and Hwansoo Lee. Iot smart home adoption: The importance of proper level automation. *Journal of Sensors*, 2018:1–11, 05 2018.

[383] L. Yang, Y. Ge, W. Li, W. Rao, and W. Shen. A home mobile healthcare system for wheelchair users. In *Proceedings of the 2014 IEEE 18th International Conference on Computer Supported Cooperative Work in Design (CSCWD)*, pages 609–614, May 2014.

[384] Y. Yang, X. Chang, J. Liu, and L. Li. Towards robust green virtual cloud data center provisioning. *IEEE Transactions on Cloud Computing*, 5(2):168–181, April 2017.

[385] Xun Yi, Athman Bouguettaya, Dimitrios Georgakopoulos, Andy Song, and Jan Willemson. Privacy protection for wireless medical sensor data. *IEEE Transactions on Dependable and Secure Computing*, 13(3):369–380, 2015.

[386] Xun Yi, Russell Paulet, and Elisa Bertino. *Homomorphic Encryption and Applications*, volume 3. Springer, 2014.

[387] Ahmed Youssef. A framework for secure healthcare systems based on big data analytics in mobile cloud computing environments. *The International Journal of Ambient Systems and Applications*, 2:1–11, 06 2014.

[388] Andrea Zanella, Nicola Bui, Angelo Castellani, Lorenzo Vangelista, and Michele Zorzi. Internet of things for smart cities. *IEEE Internet of Things Journal*, 1(1):22–32, 2014.

[389] J. K. Zao, T. T. Gan, C. K. You, S. J. R. Méndez, C. E. Chung, Y. T. Wang, T. Mullen, and T. P. Jung. Augmented brain computer interaction based on fog computing and linked data. In *2014 International Conference on Intelligent Environments*, 2014.

[390] S. Zardari, N. K. Khan, and M. A. Memon. Systematic Analysis of Risks in Cloud Architecture. *Int. J. of Computer Science and Technology (IJCST)*, 14(11), 2016.

[391] Gail-Joon Ahn, Dijiang Huang, and Shanbiao Wang. Towards temporal access control in cloud computing. In *Proceedings IEEE INFOCOM, INFOCOM 2012*, 2576–2580, 2012.

[392] H. Zhang, Y. Xiao, S. Bu, D. Niyato, R. Yu, and Z. Han. Fog computing in multi-tier data center networks: A hierarchical game approach. In *2016 IEEE International Conference on Communications (ICC)*, pages 1–6, 2016.

[393] Heng Zhang, Peng Cheng, Ling Shi, and Jiming Chen. Optimal dos attack scheduling in wireless networked control system. *IEEE Transactions on Control Systems Technology*, 24(3):843–852, 2015.

[394] J. Zhang, Z. Wang, Z. Yang, and Q. Zhang. Proximity based IoT device authentication. In *IEEE INFOCOM 2017 - IEEE Conference on Computer Communications*, pages 1–9, May 2017.

[395] Kuan Zhang, Xiaohui Liang, Mrinmoy Baura, Rongxing Lu, and Xuemin Sherman Shen. PHDA: A priority based health data aggregation with privacy preservation for cloud assisted WBANs. *Information Sciences*, 284:130–141, 2014.

[396] Kuan Zhang, Jianbing Ni, Kan Yang, Xiaohui Liang, Ju Ren, and Xuemin Sherman Shen. Security and privacy in smart city applications: Challenges and solutions. *IEEE Communications Magazine*, 55(1):122–129, 2017.

[397] Lixia Zhang, Alexander Afanasyev, Jeffrey Burke, Van Jacobson, KC Claffy, Patrick Crowley, Christos Papadopoulos, Lan Wang, and Beichuan Zhang. Named data networking. *SIGCOMM Comput. Commun. Rev.*, 44(3):66–73, July 2014.

[398] Wen Zhang, Carl A. Gunter, David Liebovitz, Jian Tian, and Bradley Malin. Role Prediction using Electronic Medical Record System Audits. *AMIA Annual Symposium Proceedings*, 2011:858–867, 2011.

[399] Y. Zhang and X. Wu. Access control in internet of things: A survey. In *arXiv*, 2016.

[400] Zhi-Kai Zhang, Michael Cheng Yi Cho, Chia-Wei Wang, Chia-Wei Hsu, Chong-Kuan Chen, and Shiuhpyng Shieh. Iot security: Ongoing challenges and research opportunities. In *2014 IEEE 7th International Conference on Service-Oriented Computing and Applications*, pages 230–234. IEEE, 2014.

[401] M. Zhanikeev. A cloud visitation platform to facilitate cloud federation and fog computing. *Computer*, 48(05), 2015.

[402] Bolei Zhou, Yiyou Sun, David Bau, and Antonio Torralba. Revisiting the importance of individual units in cnns via ablation, 2018.

[403] Jun Zhou, Zhenfu Cao, Xiaolei Dong, and Athanasios V. Vasilakos. Security and privacy for cloud-based iot: Challenges. *IEEE Communications Magazine*, 55(1):26–33, 2017.

[404] Y. Zhang, W. Zhou, and P. Liu. The Effect of IoT New Features on Security and Privacy: New Threats, Existing Solutions, and Challenges Yet to Be Solved. *IEEE Internet of Things Journal*, pages 7–17, 2018.

[405] Jan Henrik Ziegeldorf, Oscar Garcia Morchon, and Klaus Wehrle. Privacy in the internet of things: Threats and challenges. *Security and Communication Networks*, 7(12):2728–2742, 2014.

[406] SM Zobaed and Mohsen Amini Salehi. Big data in the cloud, 2019.

Index